River Cities in Asia

International Institute for Asian Studies

Publications

The International Institute for Asian Studies (IIAS) is a research and exchange platform based in Leiden, the Netherlands. Its objective is to encourage the interdisciplinary and comparative study of Asia and to promote (inter)national cooperation. IIAS focuses on the humanities and social sciences and on their interaction with other sciences. It stimulates scholarship on Asia and is instrumental in forging research networks among Asia Scholars. Its main research interests are reflected in the four book series published with Amsterdam University Press: Global Asia, Asian Cities, Asian Heritages and Humanities Across Borders (HAB) Methodologies.
IIAS acts as an international mediator, bringing together various parties in Asia and other parts of the world. The Institute works as a clearinghouse of knowledge and information. This entails activities such as providing information services, the construction and support of international networks and cooperative projects, and the organization of seminars and conferences. In this way, IIAS functions as a window on Europe for non-European scholars and contributes to the cultural rapprochement between Europe and Asia.

IIAS Publications Officer: Mary Lynn van Dijk

International Institute for Asian Studies

Publications

First published in 2022 by Amsterdam University Press Ltd.

Published 2025 by Routledge
4 Park Square, Milton Park, Abingdon, Oxon OX14 4RN
605 Third Avenue, New York, NY 10158

Routledge is an imprint of the Taylor & Francis Group, an informa business

ISBN: 9789463721851 (hbk)
ISBN: 9781041185604 (pbk)
ISBN: 9781003703129 (ebk)
NUR 740

Cover photo: Daily floating market located close to the confluence of the Barito and Kuin Rivers in Banjarmasin, South Kalimantan, Indonesia
Photographer: Vera D. Damayanti

Cover design by: Coördesign, Leiden

DOI: 10.5117/ 9789463721851

For Product Safety Concerns and Information please contact our EU representative:
GPSR@taylorandfrancis.com
Taylor & Francis Verlag GmbH, Kaufingerstraße 24, 80331 München, Germany

River Cities in Asia

Waterways in Urban Development and History

Edited by
Rita Padawangi,
Paul Rabé and
Adrian Perkasa

Routledge
Taylor & Francis Group
LONDON AND NEW YORK

Table of Contents

List of Figures and Captions

List of Tables and Captions

Acknowledgments

The *River Cities* volume was born out of a symposium hosted by Airlangga University in Surabaya, Indonesia in December 2017, entitled *"River Cities: Water Space in Urban Development and History"*. The symposium was organized by the Urban Knowledge Network Asia (UKNA) of the International Institute for Asian Studies (IIAS), in collaboration with Airlangga University, to contribute innovative ways of thinking about how to better integrate rivers, creeks and canals—including their environmental, historical, social, political, cultural and economic dimensions—into the fabric of contemporary cities.

We wish to thank the Faculty of Humanities at Airlangga University for their support and hospitality during the symposium, as well as the City and Mayor's office of Surabaya, which invited us to the City Hall for an unforgettable dinner and cultural show that made us feel so welcome in this marvelous "city of heroes" (*Kota Pahlawan*).

The UKNA and IIAS provided a generous financial contribution to make the symposium possible, for which we are very grateful. Special thanks are due to Xiaolan Lin, Assistant Coordinator of UKNA, for her help in organizing the Surabaya gathering. We also wish to thank the paper presenters and volume contributors for their patience and collaboration during the (seemingly) long years of the COVID-19 pandemic, as we worked to put the pieces together for this book and incorporated the many useful insights and improvements suggested by the peer reviewers. We are very grateful to the reviewers, who have dedicated time and effort to provide meaningful comments.

Our biggest thanks are to Mary Lynn van Dijk, IIAS Publications Officer, who worked tirelessly behind the scenes to make this entire book project possible. Without her constant and reliable follow-up, editing work, and communication with the chapter authors, the publisher and us as Editors, "River Cities" would never have materialized.

Finally, we are very pleased and thankful to note that the "River Cities" book has spawned a transdisciplinary action research network by the same name, coordinated from IIAS, which seeks to promote revitalization of rivers and the landscapes, riverscapes, cities and neighborhoods that co-exist with them. This is surely the biggest achievement that a book can hope to have, namely, to translate insights from theory and practice to help build knowledge in other places and contexts. We started our Introductory essay with the observation that "cities and water have a love-hate relationship". Let us hope that this book—and the network that flows from it—can contribute to a new relationship where love can dominate hate, based on better understanding and mutual respect.

1 River Cities in Asia

Waterways in Urban Development and History

Rita Padawangi, Paul Rabé, and Adrian Perkasa

Cities and water have a love-hate relationship (Feldman 2017). This is especially true of rivers in many cities in Asia, which, like cities in the rest of the world, owe their locations to rivers and the trading opportunities and water sources these rivers provided. In recent years, cities across China have been beautifying and extending their waterfronts, and cities as diverse as Singapore and Seoul are turning their rivers into assets as part of urban redevelopment schemes or restoring them in an effort to bring nature back to the city. Nevertheless, many other cities in Asia have their backs turned to their rivers. Where rivers were once trading and transport arteries and hubs of social and cultural life, many rivers today have suffered neglect as roads and evolving trading patterns have supplanted the rivers' economic and social functions. Their decline has been accompanied by environmental destruction, as their waters have become polluted and serve as dumping grounds for solid waste. Moreover, riverbank settlements have evolved into legally ambiguous spaces, as old settlements were detached from land formalization regimes and were subjected to environmental deterioration from the rivers. Far from being an asset, these rivers have become an eyesore—and occasionally also a threat, owing to flooding exacerbated by poor planning and a poor understanding of the place of these water bodies in the wider regional ecosystem.

This book seeks to uncover the relationship between rivers and cities from a multidisciplinary perspective in the humanities and the social sciences. The chapters in this book consist of contributions from both scholars and practitioners; all aim to contribute innovative ways of thinking about how to better integrate rivers, creeks, and canals—including their environmental, historical, social, political, cultural, and economic dimensions—into the fabric of contemporary cities.

Padawangi, Rita, Paul Rabé and Adrian Perkasa, *River Cities in Asia*.
Taylor & Francis Group, 2022
DOI: 10.5117/9789463721851_CH01

Although there are other scholarly publications on urban rivers, this book distinguishes itself by bringing together the humanities and social sciences in emphasizing history and practice-oriented ethnography, with particular attention paid to specific, local urban development issues. We seek to examine the built and natural environment dialectics "from below," embedded in time and space, countering the often top-down and ahistorical perspectives of many planning manuals.

The focus is on cities in Asia, as the region is currently home to rapid urban development, with significant intervention in the environment occurring at a large scale in compressed timeframes (Simone and Pieterse 2017). Meanwhile, governance is often subject to postcolonial conditions in which much of the hailed policy and planning interventions originate in the West. There are cases of cities in which policymakers aspire to manage everything, in contexts where they are unable to control or predict many aspects. This context, in which lack of control and inability to predict negatively affect the designing of suitable policies and projects, is often rooted in the overwhelming scales of urban development and limits on the resources required to respond to the challenges.

The Concept of the "River City"

At the heart of the book is the concept of the "river city," which has both literal and implicit meanings. While the literal meaning of the term implies river and city in a shared location, the implicit meaning encompasses proximity between the river and the city in the previously mentioned dimensions: time and space, natural and social. The existence of the city is inseparable from the river, and vice versa. Understanding the time dimension of the city goes beyond the historical aspect to include the processes in which the river defines the city and the city, the river. Spatially, the time aspect is observable in the multiple layers that connect, disconnect, and reconnect the city and the river, as well as city-river spaces that are seamlessly integrated. The natural dimensions of the river, as will be expanded further in this book, appear in many different guises, as consumable resources and as infrastructure, and as natural habitats and platforms for all manner of human activities, from worship to leisure and tourism. Finally, and very importantly, the social dimensions are reflected in the relationship between urban communities (and neighborhoods) and the river, in which the river is socially constructed and shaped by these communities, but the existence of these communities and societies is simultaneously defined by the river.

At this point, we should recognize that use of the word "city" in relation to rivers is somewhat problematic, for at least two reasons. The first is because the intense relationships between humans and rivers as described above have played out in locations where humans have settled along rivers and streams, and these locations may not always have been "cities" per se. Instead, they may have been smaller settlements, hamlets or villages, which may or may not have grown organically into denser and larger settlements ("cities") over time. Another aspect of this is that it is rare for an entire "city" to have an equal relationship with one particular river. Rather, it is probably more common for certain neighborhoods to have a more intimate relationship with a nearby river or stream than other neighborhoods in the same city. A second reason for caution in using the word "city" relates to the place of rivers as part of a system. All streams and rivers reach beyond the boundaries of cities, upstream and downstream. As such, they transit through cities and have relationships to water catchment areas upstream and delta hinterlands downstream. Thus, we can say that "cities" (or their neighborhoods) are only selected points in an entire system that influences how rivers are shaped, and in turn how rivers shape their landscapes. These landscapes are by definition socio-ecological systems that consist of natural and human-modified ecosystems, with a configuration of topography, vegetation, land use, and settlements influenced by the ecological, historical, economic, and cultural processes and activities of the area (Scherr et al. 2013). Acknowledging the presence of such a broad system enables us to contextualize the multiple uses of urban rivers, as habitats for flora and fauna, for ecosystem services, trade, transport, leisure, culture and worship, waste management, and as part of an urban political ecology.

Having mentioned all this, and having provided these qualifications, we nevertheless return to use of the (less than complete) term "city" as the focus of our human-river relationship in this book. While the term "landscape" would provide a more complete backdrop for the human-river interactions studied, it is the density of human activity inherent to "cities" and its interplay with rivers that is really the main interest of this and other chapters in this book.

Notes on Geographical Focus

The geographical focus of the book necessarily has its limitations. With its focus on four rivers in India, three rivers in Southeast Asia, and one each in Korea, Taiwan, and mainland China, the book cannot—and indeed does not—pretend to cover all of "Asia." Given the huge diversity in riverine conditions in and around cities in Asia, no single book can claim to be comprehensive on this subject. Instead, this book's focus reflects the research areas of its main contributors, who collaborated as part of the International Institute for Asian Studies (IIAS) Urban Knowledge Network Asia's River Cities project in 2017. The editors and contributors acknowledge the excellent work that has been done on other critical rivers with urban implications in other parts of South, Southeast, and East Asia, and they acknowledge that further work is needed to expand and deepen the "river cities" lens even to other areas along the same rivers, in different times and conditions, given the highly dynamic circumstances of urbanization and ecology in the regions studied.

Another limitation of this book is the relative absence of explicit discussion of the trans-nationality or trans-territoriality of rivers. Transboundary conflicts have many impacts on rivers and riverine communities—including urban ones—often pitting upstream and downstream users against each other in clashes over water use, water quality, and flood control. Some of these issues come up in several of the chapters (in the planning of flood control measures in Taipei, or in discussions of water pollution in Varanasi, for example), although in a more implicit way, in the sense that cross-boundary conflicts are not the focus of these chapters. Here, too, the editors acknowledge the important work that has been done on riverine jurisdictional issues and they recognize the need to include these political dimensions in a more explicit fashion in further research.

A legitimate question in relation to this book and its title may be, "Is there anything specifically 'Asian' about these river cities studied?" At one level, the answer would almost certainly be "No." The riverine phenomena vividly described in this book—including land use conflicts on the water's edge, informality, redevelopment, pollution and solid waste management, flooding and flood control, and river deterioration—are universal and would be instantly recognizable to readers in cities in other parts of the world. At another level, however, there are undeniably unique, local histories and characteristics to the stories recounted in this volume, from the stone temple complexes in Kanchi to the popular processions along the bridge and temple structures in Jiangnan region, and from the covering and subsequent uncovering of the river in Suwon City to the riverbank improvement schemes in Java.

A more difficult question follows from this: "Is there anything that connects all these examples as typically 'Asian'?" And are there specifically "Asian" aspects to the relationships of cities and rivers described in this book?

In the present Asian Cities book series, contributors to the three volumes of the Urban Knowledge Network Asia (UKNA)[1] engaged in many discussions about the question of whether it makes sense to talk about "Asian cities." In Volume 3, Li Shiqiao argues that there are important differences in Western and Asian urban traditions, enabling us to speak tentatively of an "Asian city" (Li 2020). According to Li, "instead of yearning for absolute freedom as in the Western city, human agency is generally understood as conditional in the Asian city" (Li 2020, 23). An Asian tradition of "immanence" (engaging thought at the level of things and events actually taking place—in contrast to Western abstract ideals of the human) has consequences for the environment: it results in a "deep humility of humans in the context of the environment ... which does not place final authority of action in humans (or their otherworldly authority in supernatural beings), but in the environment as it is" (Li 2020, 35). Li Shiqiao's thoughts are intriguing, as the implications for river cities of placing final authority of action not in humans but in the environment "as it is" are potentially enormous. However, there is certainly no unanimity on this subject in the literature. For many others, the notion of an "Asian city" is too essentialist, and cannot do justice to the huge variety of distinct cultural and political traditions across the continent (itself a disputed concept, for where are its borders?). For this reason, the UKNA prefers to use the term "cities in Asia" instead of "Asian cities."

Rather than arguing for an "Asian city" or "Asian river city," this book *does* acknowledge that there are a number of phenomena in the urban river landscape that many countries in Asia share, and which are perhaps more pronounced in this region than in others. Howard Dick, in this volume, advances three reasons that would appear to be particularly applicable to most river cities in Asia. He starts with an observation, based on history: in many parts of urban Asia, people lived beside the sea and rivers and "accepted the tides and monsoons as part of the rhythm of life." But since the mid-twentieth century, this more-or-less resilient way of life has given way to new patterns of urbanization, less attuned to the environment. The result is that Asia's cities are increasingly caught in a "tightening three-way vice."

> First, population growth, development and deforestation in the hinterland has increased water run-off and downstream siltation. Second, rapid urbanization combined with modernization in the form of paved roads and brick and concrete buildings has created heat islands, reduced the

1 Volume 1: *Ideas of the City in Asian Settings* (2019); Volume 2: *Cities in Asia by and for the People* (2019); and Volume 3: *Future Challenges of Cities in Asia* (2020).

> absorption of urban surfaces, and in some cities led to irreversible subsidence. Third, climate change is leading to an irreversible trend of rising sea levels while increasing the intensity of storms. (Howard Dick, this volume)

Dick suggests that the "tightening three-way vice" may be understood from the perspective of biological systems. Technological advances and rising living standards have meant that cities in Asia have accommodated millions of additional people in a relatively short time-span. However, coastal and riverine cities will not be able to sustain these demographic and economic pressures unless they adapt, "with more sophistication than the present-day concrete, high-energy jungles," to a changing natural environment. If not, then an organism that "expands without limit will eventually reach a threshold at which it ceases to be environmentally sustainable, with potentially catastrophic results." It is this grim reality that we suggest today connects many of the river cities in Asia featured in this volume: historical and climatological trends that are not unique to Asia, but which arguably are more pronounced for coastal and riverine cities in this part of the world than any other region, due to the unique combination of large population densities and the rapid pace of urban growth in locations that are extremely vulnerable to climate change.

Structure of the Book

Explorations of various localities in this volume go deeper into multiple dimensions rather than the book being just another collection of case studies. Much of urban studies literature, particularly that which deals with contemporary issues that require policy and planning interventions, tends to pull together case studies from various localities to search for common challenges and best practices. While these comparisons may be useful to a certain extent, the approach suffers from the lack of contextualization to place and time. As a result, many policy and planning interventions in contemporary river cities in Asia import best practices from elsewhere that may work in one aspect but not in others, and necessarily involve trade-offs. We argue that the so-called trade-off to justify pragmatic policy and planning approaches is problematic, as it easily causes other social and environmental problems in exchange for project-based interventions. As observed in many of the chapters in this book, river interventions in the city are always inseparable from revenue generation, production of space, and political gains, while in many cases they also use social and environmental pretexts to normalize the marginalization of weaker social-political groups as development collateral. This book challenges the simplification of river city approaches to advocate for more attention to contextualizing rivers and finding solutions within the relevance of the historical, social, and cultural contexts of the city, particularly in the relationships between cities and their rivers.

Consequently, the chapters in this book are organized according to four categories of investigation: (1) Rivers and cities in historical perspective; (2) Neighborhoods and social life of riverine communities; (3) Riverfront and riverbank settlement and design interventions in Asia; and (4) Urban policy perspectives and innovations.

Rivers and Cities in Historical Perspective

The emergence of urban settlements is inseparable from the availability of resources, of which rivers provide many. "Water is the lifeblood of any city ... [as] a supply of water while simultaneously resolving the issues of provisioning, trade, transport, and defense," writes Howard Dick (this volume). While many cities across the globe are situated along water bodies (UN-HABITAT 2006), many cities in Asia are associated particularly with rivers or other fresh water sources. The ancient city-kingdom of Angkor (ninth century) has an inland geography that relied on an extensive water system, and so was the capital of Lanna (currently northern Thailand) that was located adjacent to the Ping and Wang Rivers. Several kingdom-capitals of archipelagic Southeast Asia, such as Trowulan, Majapahit, relied on river systems both as a source of fresh water and for transportation.

Water (including rivers) has had many important functions in Asia and particularly Southeast Asia's historical development. The dynamics of Southeast Asia's history—both on the mainland and in its islands—can easily be found along its rivers, which provided access to inland markets and sources of commodities (Sutherland 2007). These rivers became a decisive factor not only in Hindu-Buddhist kingdoms but also later in the colonial era. The Bengawan Solo River in Java is one such case, which Nicholas Hartingh, the Dutch East-Indies Company (*Verenigde Oost-Indische Compagnie*) Governor of North-Eastern Java (1746–1761), viewed as a favorable feature of Gresik in East Java. The King of Surabaya granted permission to the Dutch to locate their trading post in Gresik, a location of Bengawan Solo's outfall to the sea, which eventually became the transportation route for commodities from agrarian areas in the interior of Java.

Rivers have been an integral aspect of socio-cultural and religious life across Asia. For example, the River Ganga's religious importance has been inscribed through centuries of traditions, practices, and cultural inscriptions. Associated with gods and goddesses, religious devotees believe in the river's power of purification, and this belief affects the ways in which the population organizes urban spaces. Historically, temples have been built along rivers in South India as well as in China. In mainland Southeast Asia, water—including baray (lakes), canals, and moats—was an enormously important part of the design of Khmer temples and sanctuaries (Stubbs 1998). In archipelagic Southeast Asia, the Brantas River in East Java was the main river of the Singasari kingdom and afterwards Majapahit, which was the biggest empire in Southeast Asia in the fourteenth century. Consequently, many myths and folklores appeared on and along the history of the river (Basundoro 2017).

The economic importance of rivers in Southeast Asia decreased soon after the introduction of roads, followed by railways. Java and Sumatra are examples of such decline. The Dutch colonial government, under Governor General Herman Willem Daendels, built the first major road in Java in 1808–1811. Despite Daendels's intention to build this road only for military motives, the Grote Postweg (the Great Post Road) in Java gradually changed the modes of transportation. By 1890 the road network in Java covered many cities and a few decades earlier, the first railroad in Java was opened. These new modes of transport replaced the old ones, such as vessels and junks. It also changed the transportation pattern on Java from north-south to east-west (Nas and Pratiwo 2002). Similar developments also occurred in Sumatra, which experienced spatial, economic, and ecological changes after the introduction of modern land transportation. Before the colonial government constructed new roads, indigenous Sumatrans in Middle Sumatra preferred waterways to overland transportation. This was reasonable because the price of transport over water at the time was more affordable than land transportation. However, new road developments affected waterways, as the removal of many plants and trees caused soil erosion at a rate of around 11–13 tons per hectare per month (Colombijn 2005).

Historical perspectives on rivers and cities do not only tell us something about the past. Rather, historical perspectives contextualize the present along a time trajectory. Historical perspectives provide alternative means to generate in-depth understanding of the relationship between rivers and city life. Consequently they help to rethink policies and project interventions on urban rivers. Therefore, historical perspectives help us not merely to return to the past, but also to better understand the present conditions for possible future interventions.

Neighborhoods and Social Life of Riverine Communities

Settlement formation along rivers has been documented in historiographies of cities around the world, including Asia. Howard Dick (this volume) notes the rise of Surabaya, one of the main coastal cities in Java, as a control point for the main port of the Majapahit capital, Trowulan, which was connected to the Brantas River system. Surabaya's economy thrived with the rise of trade and commerce as well as religious crusades through ports. The role of commerce in river settlement formation was also prominent in the case of Chennai's Buckingham Canal (Coelho, this volume), which facilitated freight transportation in the 1800s, during the British colonization of India. River cities that thrive on trade and commerce, therefore, are likely to be coastal port cities where rivers connect these cities to their hinterlands.

Social practices relate to the particular geographies where inhabitants' settlements are located. The life of these settlements, therefore, is not only about the pragmatic functions of the river, such as provision of water and transportation, but also about the ways in which rivers intersect with everyday life. The emergence of riverside temple networks in South India is an example of how religious beliefs transect social practices and services, as temples also functioned as important municipal institutions, such as hospitals, schools, and providers of social assistance, at least in the eleventh century (Stein, this volume). The close relationship between temples and rivers in South China made rivers function as pathways for religious processions, and also as markets on non-procession days (Zhao, this volume). Riverine communities are spatially compelled to integrate rivers with their lives and livelihoods, as reflected in mundane practices such as drinking and washing, recreational activities, and transcendental practices that elevate the symbolic meaning of the river.

There are different variations of integration of river settlements with the river as water and space. First is the use of water space as recreational space shared with friends and family in the neighborhood (Westhues 2017). Through this usage, river water functions as a public space that brings people together, and at the same time may also allow different groups to use the space differently. Second is the use of water space as infrastructure, which is also a form of public use but more immersed in the fulfillment of daily basic needs. This relates to the use of water spaces as transportation, to get from one point to another along the waterway, as well as the use of water itself to be consumed. Third is the use of water space integrated with private spaces, in the form of floating houses, for example, as shown in the chapter by Vera Damayanti (this volume). This conversion into private spaces is also inseparable from the role of the river as the main transportation route that includes economic and political activities. Fourth is the symbolic use of water space, as river water may carry with it mystical and mythical powers, representing the relationship between the social and the natural world (Westhues 2017).

As manufacturing industries become the main drivers of urban growth, landscapes change. The water retention function of water catchment areas often has to compete with other land uses—such as industry and housing—that run counter to retaining water owing to their reliance on hard surfaces. While earlier versions of riverine settlements also encroached on water catchment areas, the mechanization of production and market expansion has significantly altered the scale of land use changes. Many cities in Asia experienced exponential growth after the 1960s, when offshoring of industrial production from the Northern hemisphere spurred rapid industrialization in cities and new industrial zones and induced economic growth and construction projects in the urban core. The subsequent population increase put strains on existing infrastructures and environments, including rivers and their catchment areas. Furthermore, the increasing reliance on land-based transportation reduced the role of the river as the main transportation spine in many cities, and in many cases turned rivers into backwaters and dump sites for sewage and waste.

Environmental degradation of rivers is impacting riverine settlements and the relationship of humans to water, and in the process, disconnecting cities and their inhabitants from the rivers. For example, pollution in the Bishnumati River, which is a result of development planning in Kathmandu ignoring natural ecological systems, has led to cultural rites and rituals being conducted inside houses rather than along the riverbanks, as was previously the case. Moreover, bottled mineral water is replacing water for drinking as well as water for rituals, as river water becomes polluted. Another case is the environmental degradation of the Ganga, the most sacred river for Hindus, which is now turning into a "toxic sludge" due to pollution and excessive usage (Siddiqui 2017). The alarming level of pollution is in stark contrast to the tens of thousands who come to the ghats of the Ganga every day to take their holy dip.

Yet another dimension of human-induced degradation of rivers is provided by the effects of climate change. Results from climate models show that upstream ice reserves for major Asian rivers—including the Indus, Ganga, Brahmaputra, Yangtze and Yellow rivers—are likely to be affected substantially by climate change, with a potentially profound effect on millions of people downstream in terms of their water supply and food security. While the precise effects are difficult to generalize across basins, these models predict that the effects in the Indus and Brahmaputra basins, in particular, are likely to be severe owing to the large populations in these basins and their high dependence on irrigated agriculture and meltwater (Immerzeel et al. 2010). River basins in other parts of Asia are also likely to be increasingly affected by climate change, as Howard Dick illustrates in this volume in the case of Surabaya, through a combination of heavier monsoonal rains and more severe floodwaters from local river catchments. Coastal cities are particularly vulnerable, with rising sea levels an additional risk factor. Excessive land development in the wrong places—often encouraged by local development policy—exacerbates the risks of flooding, through land subsidence and the paving over of natural wetlands and reservoirs. The most risk-prone areas are often poor urban settlements, where residents lack the means to live in safer areas or to upgrade their houses, or lack the tenure security to invest in improvements to housing quality that reduce vulnerability to natural hazards (Mitchell et al. 2015).

The relationships between settlements and rivers are not static and evolve along with urban society. Given the historical and socio-cultural understandings of river cities, how would the river city concept be useful in responding to contemporary urbanization challenges? The embeddedness of rivers in city life is inseparable from the rivers' ability to provide resources for urban settlers and processes, but many urban river systems are environmentally degraded (Findlay and Taylor 2006; Rademacher 2009; Taylor 2015). The following sections take a step further in attempting to relate the humanities and social sciences to the current environmental and social issues of urban rivers, and in describing how the river city perspective contextualizes conceptual and policy actions in addressing urbanization challenges.

Riverfront and Riverbank Settlement and Design Interventions in Asia

Being "excellent indicators of environmental change" (Nilsson et al. 2017, 1), urban rivers highlight environmental problems when they degrade. Environmental degradation of urban river systems is a direct consequence of rivers being treated opportunistically as resources to be extracted, be it the water, stream, or catchment area. In a study on the politics of environmental and water pollution in East Java, Lucas and Djati (2007, 322) pointed to the role of factories that were developed along the Surabaya River in the 1970s without proper waste treatment plants in increasing pollution on this tributary of the Brantas River. The pollution was so severe that a popular

Javanese jingle grew out of it: "when the dog is dead, throw it into the river, when the river is flooding, leave the dog on the bank." The case of Surabaya River pollution exemplifies the trajectory across many cities in rapidly industrializing countries in Asia and beyond (Williams 2001; Dobbs 2002; Selman et al. 2010; Padawangi 2012; Padawangi and Douglass 2015).

Together with pollution, environmental degradation of water catchment areas adds to the threat of flooding, which is currently the most common disaster to hit urban areas (Douglass 2013; Liao 2012). Changing land use that reduces water retention increases run-off to built-up areas, often downstream from the places where the land use change occurs, beyond city boundaries and sometimes even national boundaries (Lu et al. 2015; Vollmer et al. 2015). Urbanization affects the entirety of river ecosystem processes and eventually affects social life in the city through ecosystem impacts, often in the form of magnifying socio-spatial problems such as segregation and discrimination (Paul and Meyer 2001; Padawangi and Douglass 2015).

Interventions in the name of reducing environmental degradation become increasingly common as rivers become sources of flooding and no longer provide water services reliably. The viability of these interventions to address the problems remain uncertain and may often worsen those problems. Design interventions often rely on reengineering rivers to become embanked canals, such as observed by Coelho in her chapter on water edge urbanisms in Chennai. Attempts towards such interventions continue to take place in other cities in Asia, as Youngah Guahk demonstrates in her chapter on Seoul. Many of these visions are results of ignorance of historical river-city connections, but they are also manifestations of ambitions toward achieving "globalized norms about green building standards and other sustainable infrastructure" that perpetuate the "global cultural currency of mobile elites" (Simone and Pieterse 2017, 40).

Coupled with the mainstream development financing paradigms that emphasize competitiveness of cities and the pursuit to achieve global city status, design interventions in Asia's river cities embrace the paradigm of putting the river back at the center of the city's cultural space, but at the same time decouple the river from self-built settlements. Rather than relating the river to evolving riverine settlements, many river restoration projects involve uprooting existing settlements to put in new infrastructure and large-scale spatial arrangements that are seemingly more environmentally friendly. The top-down intervention on the Sabarmati River in Ahmedabad, for example, is a mega project that claimed to give a new face to the city through environmental improvement, social infrastructure delivery, and sustainable development, but the population was displaced and natural riverbanks were replaced by high concrete retaining walls (Parthasarathy 2017). Another example is the Ciliwung River in Jakarta, on which the provincial government fetishized concrete embankments as the solution to flooding, and hence justified the forced evictions of riverine communities who were accused of being river polluters.

The consequence of these top-down design interventions that are obsessed with concrete embankments and mega-scale projects is the further separation between the river and the city. Echoing Lefebvre's claim that "nature has been defeated," (1991, 31) those that are still allowed to exist in the city are restricted to engaging in leisure activities, which is again a function designed to accumulate wealth through space, without having to anchor anyone into a particular space and place for the long term. River transportation may return but is possibly reduced to tourist boats for leisurely sightseeing rather than as a professional linkage between the society and the river water. Examples of such transformation are abundant, including the Love River gondolas in Kaohsiung, Taiwan and the bumboats that are stationed at Singapore River (Savage et al. 2004). Many visions for river improvements—such as images for the Pasig River in Metro Manila, the Ciliwung River in Jakarta, the Red River in Hanoi, and the Han River project in Seoul that was eventually held back when the city administration changed—emphasize tourism in their development projects. While tourism can provide some opportunities, it also perpetuates the forces and processes of the production of space, namely wealth accumulation and economic class injustices, which had induced the environmental degradation of the river and the river city in the first place.

In current global water policy discourse there are several alternative approaches that aim to better reconcile cities with their rivers. Shannon (2008) promotes the "water urbanism" concept to allow water to have its space instead of imposing tight control through hard-surface projects. Using the term "soft engineering" to describe water urbanism, the concept refers to the (re-)integration of community life with water. However, much of the current practices in crafting design-based solutions to river problems are based on descriptive readings of landscapes with insufficient depth of thinking and examination into the social and cultural geographies of the place. In tandem with the advances of visualization technologies, descriptive and superficial design ideas may claim superiority in policy landscapes, and may sustain social and economic systems that are counter to environmental sustainability (Halpern 2014).

The Integrated Water Resource Management (IWRM) approach is one that is promoted by governments and multiple donor organizations as a comprehensive approach that involves stakeholders while fostering economic growth (Hassing et al. 2009). It has also been touted as promoting the democratic governance of water resources and as improving human health (GWP 2011). While it has been widely embraced in many countries and transboundary river planning initiatives, IWRM has also received criticism from scholars for both its conceptualization and implementation. The application has been subject to institutional barriers as well as superficial "compilation of donor-funded projects" because there has been lack of agreement among IWRM proponents as to what aspects should be integrated (Suhardiman et al. 2015, 285), while conceptually it does not sufficiently recognize the political nature of water resources management, hence the uneven power landscape among different actors involved. The rural-urban divide is an obvious power inequality within the river basin, especially when economic growth is the main agenda: cities, in growth-centered developmentalist thinking, are engines of economic growth, and are more powerful in gaining legitimacy to degrade their rivers to achieve that growth. Consequently, while IWRM is widely promoted in various river basins and transboundary river management initiatives, many IWRM cases are outside city boundaries, and concrete-dominant design interventions and green cosmetic landscapes for leisure-oriented mega-projects are still the mainstream language in urban river rehabilitations.

Another alternative to the engineering-only approach to river basin management is "room for the river," which has been pioneered in the Netherlands during the past two decades as part of a large government-funded program, and which is now gaining a foothold in water management policies in other countries around the world, including in Asia. "Room for the river" is based on the premise that seeking more and more "control" of water—through ever-higher dikes, for example—is doomed to failure given the vagaries of climate change. Instead, long-term safety may be better achieved by introducing nature-based principles and giving rivers more room to flood from time to time. In the transition presented by "room for the river," the traditional engineering and control paradigm in flood risk management is gradually being replaced by an integrated approach that incorporates various other disciplines, such as water management, spatial planning, and ecological principles (Rijke et al. 2012). In this spirit, the Dutch government has funded over 30 projects along major rivers, including in urban areas, in partnership with local governments and following lengthy planning sessions with local communities. But while "room for the river" brings nature back in to the policy calculus, it still represents a program very much designed by experts. Moreover, it is not clear whether the program will be expanded beyond the first generation of projects, as the construction lobby fights to reclaim the right to build infrastructure to rein in rivers in other areas not covered by the program.

Urban Policy Perspectives and Innovations

In spite of the continued dominance of hard infrastructure paradigms that have massively affected many river cities, thereby degrading the relationship between the river and the city, there have been several innovations to improve the conditions of riverbanks as well as river settlements. These innovations emphasized not reversing history, but redefining contemporary river cities. As previously mentioned, historical perspectives aim not to trap cities in the past, but rather to contextualize the analysis of current situations in order to inform decisions for the future. Contemporary urban conditions bring new challenges because of the unprecedented pressure of human population needs on river systems as well as the necessary adaptations to global climate change (Nilsson et al. 2007). While a return to pristine ecosystem conditions is unrealistic, and each river may have its specific contexts, the consideration of the four distinct dimensions of time, space, natural, and social are crucial.

There are at least three themes emanating from contemporary publications advocating for innovative interventions and policies. First is a multidisciplinary approach that is often neglected in river restoration and rehabilitation plans, as demonstrated by projects that reduce urban rivers to beautiful landscapes while marginalizing prior riverine settlements. The compartmentalization of expertise in relation to rivers often becomes a hindrance to achieving a comprehensive understanding of river system dynamics. Even when different disciplines join forces to design river interventions and policies, interactions among these disciplines are also an important factor, as an iterative process is crucial in integrating various perspectives (Vollmer et al. 2015).

The second is an emphasis on community involvement. In observations of five rivers in three different cities in Indonesia, Prescott (this volume) identified community involvement as an important socio-cultural factor, and transformations are often triggered from the "ground up." Also from observations in Indonesia, Taylor (2015, 634) argues that it is important to "provide tangible results that are meaningful to people" and to do that it is important to adopt "inclusive and participatory governance approaches" that have been proven to result in "more successful resilience outcomes." Public engagement can also lead to organizational, institutional, and individual learning through deliberative engagement, although the learning levels, processes, and paces vary (Petts 2007). It should be noted, though, that there are cases in which transformations claim to be participatory but may not fully represent different segments of society. These pseudo-participatory approaches occur for various reasons, spanning various scales of power inequality—including in the smallest neighborhoods—and populist agendas of policymakers (Padawangi 2018).

The third theme is the importance of spatial justice in any river policy and project innovations, as political powers are unevenly spread in the landscapes of urban rivers. The river city concept presumes meaningful relationships among spaces, environments, and inhabitants of rivers and cities, but these meaningful relationships presume particular attention be given to populations that are often most marginalized. The case of Shezidao raises questions of spatial injustice as the government excluded it from the flood control plan (Liao, this volume). Furthermore, the impact of changing climate on rivers affects various groups of population unevenly in the city and beyond. "The people who are vulnerable to climate change are often the urban poor living and working in situations of physical *and* social vulnerability; as such, their needs are substantial and immediate" (Taylor 2015, 634).

River cities will be the first to bear the impacts of climate change, and therefore urgently need to innovate. The current trajectory of using river rehabilitation as a theme for beautification projects is not the answer to contemporary environmental and social challenges that have been accumulating over many centuries and have exacerbated exponentially in the recent years of river city histories. Moreover, as Howard Dick, in this volume, writes:

> More investment in the clumsy, environmentally damaging infrastructure of roads, parking and cheap energy-inefficient buildings is not a solution. Twentieth-century technologies of consumption and investment are not sustainable without environmental adaptation.

Furthermore, marginalizing acts such as forced evictions to clear river lands that further decontextualize the city from its time-space, nature-social continuum, take us further from the river city concept. Innovative policies, programs, and projects to rehabilitate rivers would bring these four dimensions together in a more synergistic fashion. "River cities," therefore, are places where these four dimensions collide and define the socio-environmental characteristics of the city, through water spaces that also function as historical, social, and cultural spaces.

Conclusion

> [A]s urban space is incorporated into logics of accumulation and use that are seldom attentive to past histories and valuation, the specificity of transformations under way also becomes occluded, in the sense what takes place "here" takes place "everywhere," something beyond local collective understanding. (Simone 2014, 120)

The discussion on river cities, starting with histories, followed by neighborhoods, interventions, and innovations, highlights the contribution of the humanities and social sciences to the subject. The quote above, from AbdouMaliq Simone's book *Jakarta: Drawing the City Near* (2014), points to the necessity to traverse disciplines in maintaining collective understanding of the city. The collective urban consciousness of river-city relations is continuously constructed by various layers of history. However, recent drives to produce urban spaces for wealth accumulation increasingly obscure historical contexts, resulting in urban river mega-projects that are designed to project a city image to the world rather than to represent the immediate socio-historical and cultural relevance.

There have been many calls for a multidisciplinary approach to understanding the role of rivers in cities, but the sense of urgency in addressing current problems often subjects proposed solutions to path-dependent sectoral actions (Vollmer et al. 2015; Findlay and Taylor 2006). What has been missing in the current literature is, first, an understanding of the city as a "river city," which means that the river system—including creeks, canals, lakes, and wetlands—is an integral part of the city in all its historical, social, cultural, and political dimensions. There have been attempts to make the case for urban ecosystems as an interdisciplinary perspective to deepen understandings of the relations between humans and their environment, and recommendations have been made for iterative processes in the integration of these various disciplines, and discussions on using technologies as tools to connect various scales of river spaces (Francis et al. 2011; Vollmer et al. 2015; Padawangi et al. 2016). However, none of these attempts have represented the mainstream development trajectory. These integrations and iterations are embedded in an understanding of the city as a river city, as the connection between the river and the city in multiple aspects of urban life demand simultaneous attention without one aspect being sacrificed for another.

There is no denying that many urban rivers are in severe need of intervention due to their high degree of environmental degradation. In light of the globally changing climate, which compounds disaster threats, it is clear that the business-as-usual approach that aims to control rivers will always be insufficient, and from a wider perspective will be even more detrimental to the environment. The humanities and social sciences are of key importance in scholarly discussions as well as in the formulating of policies, programs, and projects. River degradation is a representation of the degradation of the city as a whole, and rehabilitating the river requires a recalibration of river-city connections to re-enable city dwellers to live with the river and reconstruct river water spaces as city life spaces. Beyond traversing multiple disciplines, this requires readiness to work at various scales, from the river basin to neighborhoods and vice versa, as a continuous reconstruction of the river city environmentally, economically, and socially. Furthermore, rehabilitating the river often requires the negotiation of administrative boundaries that, in many cases, work against a holistic view of the river.

Changing the development paradigm that currently dominates urban river rehabilitation scenarios is a long-term process that would not be achievable just through a single book, but we hope this volume will be part of a broader effort to rejuvenate river cities. Particularly in Asia, in which the sense of urgency emanates through the sheer scale of urbanization that changes the continental landscape, the river cities of the past, of today, and of the future cannot be taken for granted, but are contexts and visions that continue to evolve in uneven landscapes of power.

References

Basundoro, P. 2017. "Myths and Folklore in the Communities along Brantas River Basin." Paper presented at *River Cities: Water Space in Urban Development and History Symposium*, Surabaya, Indonesia, 11–12 December [Unpublished paper].

Colombijn, Freek. 2005. "A Moving History of Middle Sumatra, 1600–1870." *Modern Asian Studies*, 39(1): 1–38.

Dobbs, Stephen. 2002. "Urban Redevelopment and the Forced Eviction of Lighters from the Singapore River." *Singapore Journal of Tropical Geography*, 23: 288–310.

Douglass, Mike. 2013. "The Urban Transition of Environmental Disaster Governance in Asia." *Asia Research Institute Working Paper Series No. 210*. Singapore.

Feldman, David Lewis. 2017. *The Water-Sustainable City: Science, Policy and Practice*. Cheltenham, UK: Edward Elgar.

Findlay, Sophia Jane, and Mark Patrick Taylor. 2006. "Why Rehabilitate Urban River Systems?" *Area*, 38(3): 312–325.

Francis, Robert A., Jamie Lorimer, and Mike Raco. 2011. "Urban Ecosystems as 'Natural' Homes for Biogeographical Boundary Crossings." *Transactions of the Institute of British Geographers, New Series*, 37(2): 183–190.

GWP (Global Water Partnership). 2011. "What is IWRM?" Available at https://www.gwp.org/en/GWP-CEE/about/why/what-is-iwrm/, accessed 5 March 2018.

Halpern, Orit. 2014. *Beautiful Data: A History of Vision and Reason since 1945*. Durham, NC and London: Duke University Press.

Hassing, Jan, Niels Ipsen, Torkil Jønch Clausen, Henrik Larsen, and Palle Lindgaard-Jørgensen. 2009. *Integrated Water Resources Management in Action*. Paris: UNESCO.

Immerzeel, Walter W., Ludovicus P.H. Van Beek, and Marc F.P. Bierkens. 2010. "Climate Change Will Affect the Asian Water Towers." *Science*, 328(5984): 1382–1385.

Lefebvre, Henri. 1991. *The Production of Space*. Oxford and Cambridge: Blackwell.

Li, Shiqiao. 2020. "Human Agency in the Asian City." In *Future Challenges of Cities in Asia*, edited by Gregory Bracken, Paul Rabé, R. Parthasarathy, Neha Sami, and Bing Zhang, 23–38. Amsterdam: Amsterdam University Press. doi: 10.5117/9789463728812_ch02.

Liao Kuei-Hsien. 2012. "A Theory on Urban Resilience to Floods: A Basis for Alternative Planning Practices." *Ecology and Society*, 17(4): 48.

Lu, Xi Xi, Chanta Oeurng, Thi Phuong Quynh Le, and Duong Thi Thuy. 2015. "Sediment Budget as Affected by Construction of a Sequence of Dams in the Lower Red River, Viet Nam." *Geomorphology*, 248: 125–133.

Lucas, Anton, and Arief W. Djati. 2007. "The Politics of Environmental and Water Pollution in East Java." In *A World of Water: Rain, Rivers and Seas in Southeast Asian Histories*, edited by Peter Boomgaard, 321–352. Leiden: Brill.

Mitchell, David, Stig Enemark, and Paul Van der Molen. 2015. "Climate Resilient Urban Development: Why Responsible Land Governance Is Important." *Land Use Policy*, 48: 190–198.

Nas, Peter J.M., and Pratiwo. 2002. "Java en De Groote Postweg, La Grande Route, the Great Mail Road, Jalan Raya Pos." *Bijdragen tot de Taal-, Land-, en Volkenkunde; On the Road: The Social Impact of New Roads in Southeast Asia*, 158(4): 707–725.

Nilsson, Christer, Roland Jansson, Björn Malmqvist, and Robert J. Naiman. 2007. "Restoring Riverine Landscapes: The Challenge of Identifying Priorities, Reference States, and Techniques." *Ecology and Society*, 12(1): 16.

Padawangi, Rita. 2012. "Climate Change and the North Coast of Jakarta: Environmental Justice and the Social Construction of Space in Urban Poor Communities." In *Urban Areas and Global Climate Change* (Research in Urban Sociology, vol. 12), edited by W.G. Holt, 321–339. Bingley, UK: Emerald.

Padawangi, Rita. 2018. "Urban Activism and Civic Space in Post-Political Era: State Actors-Civil Society Relations in Jakarta and Surabaya." Paper presented at

(Re)conceptualising Asian Civil Society in the Age of Post-Politics, 16–17 August. Singapore: Asia Research Institute, National University of Singapore.

Padawangi, Rita, and Mike Douglass. 2015. "Water, Water Everywhere: Toward Participatory Solutions to Chronic Urban Flooding in Jakarta." *Pacific Affairs*. 88(3): 517–550. doi:10.5509/2015883517.

Padawangi, Rita, Etienne Turpin, Herlily, Michaela F. Prescott, Ivana Lee, and Ariel Shepherd. 2016. "Mapping an Alternative Community River: The Case of the Ciliwung." *Sustainable Cities and Society*, 20: 147–157.

Parthasarathy, Rengarajan. 2017. "Narmada Water on Sabarmati River: A Ship of Theseus?" Paper presented at *River Cities: Water Space in Urban Development and History Symposium*, Surabaya, Indonesia, 11–12 December [Unpublished paper].

Paul, Michael J., and Judy L. Meyer. 2001. "Streams in the Urban Landscape." *Annual Review of Ecology and Systematics*, 32: 333–365.

Petts, Judith. 2007. "Learning about Learning: Lessons from Public Engagement and Deliberation on Urban River Restoration." *The Geographical Journal*, 173(4): 300–311.

Rademacher, Anne. 2009. "When Is Housing an Environmental Problem? Reforming Informality in Kathmandu." *Current Anthropology*, 50(4): 513–533.

Rijke, Jeroen, Sebastiaan van Herk, Chris Zevenbergen, and Richard Ashley. 2012. "Room for the River: Delivering Integrated River Basin Management in the Netherlands." *International Journal of River Basin Management*, 10(4): 369–382.

Savage, Victor R., Shirlena Huang, and T.C. Chang. 2004. "The Singapore River Thematic Zone: Sustainable Tourism in an Urban Context." *The Geographical Journal*, 170(3): 212–225.

Scherr, Sara J., Seth Shames, and Rachel Friedman. 2013. "Defining Integrated Landscape Management for Policy Makers." *EcoAgriculture Policy Focus*, 10: 1–6.

Selman, Paul, Claudia Carter, Anna Lawrence, and Clare Morgan. 2010. "Reconnecting with a Recovering River through Imaginative Engagement." *Ecology and Society*, 15(3): 18.

Shannon, Kelly. 2008. *Water Urbanisms*. Amsterdam: SUN.

Siddiqui, Danish. 2017. "Dying 'Mother Ganga': India's Holy River Succumbs to Pollution." *Reuters*, 10 July. Available at https://www.reuters.com/article/us-india-ganges/dying-mother-ganga-indias-holy-river-succumbs-to-pollution-idUSKBN19V0OG, accessed 20 November 2018.

Simone, AbdouMaliq. 2014. *Jakarta: Drawing the City Near*. Minneapolis, MN: University of Minnesota Press.

Simone, AbdouMaliq, and Edgar Pieterse. 2017. *New Urban Worlds: Inhabiting Dissonant Times*. Cambridge, UK: Polity Press.

Stubbs, John. 1998. "Water at Angkor: More to Wonder About." *Orientations*, 29(1): 56–59.

Suhardiman, D., Floriane Clement, and Luna Bharati. 2015. "Integrated Water Resources Management in Nepal: Key Stakeholders' Perceptions and Lessons Learned." *International Journal of Water Resources Development*, 31(2): 284–300.

Sutherland, Heather. 2007. "Geography as Destiny? The Role of Water in Southeast Asian History." In *A World of Water: Rain, Rivers and Seas in Southeast Asian Histories*, edited by Peter Boomgaard, 27–70. Leiden: Brill.

Taylor, John. 2015. "A Tale of Two Cities: Comparing Alternative Approaches to Reducing the Vulnerability of Riverbank Communities in Two Indonesian Cities." *Environment and Urbanization*, 27(2): 621–636.

UN-HABITAT. 2006. *State of the World's Cities 2006/7*. London: UN-HABITAT and Earthscan.

Vollmer, Derek, Diogo Costa, Ervine Lin, Yazid Ninsalam, Kashif Shaad, Michaela Prescott, Senthil Gurusamy, Federica Remondi, Rita Padawangi, Paolo Burlando, Christophe Girot, Adrienne Grêt-Regamey, and Joerg Rekittke. 2015. "Changing the Course of Rivers in an Asian City: Linking Landscapes to Human Benefits through Iterative Modeling and Design." *Journal of the American Water Resources Association (JAWRA)*, 51: 672–688.

Westhues, Kay. 2017. "Flowing Well Park: Localized Water Knowledge, Community, and Stewardship as Expressed in Community Springs." *Southeastern Geographer*, 57(2): 107–109.

Williams, Brett. 2001. "A River Runs through Us." *American Anthropologist*, 103(2): 409–431.

About the Editors

Rita Padawangi is Associate Professor at the Singapore University of Social Sciences. She received her PhD in Sociology from Loyola University Chicago. Her research interests include participatory urban development and social movements. She is co-coordinating the collaborative Southeast Asia Neighborhoods Network (SEANNET) and is the editor of the *Routledge Handbook of Urbanization in Southeast Asia* (2019).

Paul Rabé is Academic Coordinator of the Urban Knowledge Network Asia (UKNA), based at the International Institute for Asian Studies in Leiden. He is also Senior Land Expert at the Institute for Housing and Urban Development Studies at Erasmus University Rotterdam, as well as the editor of the Asian Cities Series.

Adrian Perkasa is a PhD candidate at Leiden University, and is a Local Principal Investigator in Surabaya as part of the Southeast Asian Neighborhoods Network (SEANNET) under the auspices of the International Institute for Asian Studies (IIAS) in Leiden. He is also a lecturer in the Department of History, Faculty of Humanities, Universitas Airlangga, Indonesia.

2 Water World to Inundation

River Cities in Southeast Asia, from Old to New Millennium

Howard Dick

Abstract
For millennia cities have thrived on rivers that supply water while enabling provisioning, trade, transport, and defense. Yet since the mid-nineteenth century new technologies have transcended the need for waterways, which instead have become obstacles. Over the same period, the built urban environment has massively expanded its hard footprint and become more vulnerable to the intrusion of water. Climate change is now giving rise to a new challenge of sustainability. This problematization is explored in the case of Surabaya, then applied to other river cities in Indonesia and elsewhere in mainland Southeast Asia. The emphasis is upon how the urban system has adapted, failed to adapt, or mal-adapted to the exigencies of a monsoonal climate and global warming.

Keywords: Surabaya, urban habitat, transport systems, flood control, climate change, Southeast Asia

Introduction

Water is the lifeblood of any city. For millennia cities have thrived on the banks of rivers, large and small, that have guaranteed a supply of water while simultaneously resolving the issues of provisioning, trade, transport, and defense. Yet "River Cities" is hardly an analytical category. There are no scientifically agreed criteria by which a flowing water source is to be categorized as a river. It is simply a matter of common usage across languages and localities. A river may be permanent or transient, fresh or tidal, depending upon weather and season; in its urban aspect it may be re-engineered as a canal or port. Here I will take "river city" in its common sense meaning

Padawangi, Rita, Paul Rabé and Adrian Perkasa, *River Cities in Asia*.
Taylor & Francis Group, 2022
DOI: 10.5117/9789463721851_CH02

as an open category that can be interpreted broadly or narrowly according to the purpose at hand.

Rivers are part of the urban landscape and often intrinsic to urban identity. Some cities have kept their rivers in a semi-natural state, some have relegated them to drains or even sewers. In Europe, North America, and Australia, de-industrialization and the shift of ports to deeper water downstream have created new opportunities for riverbank land to be recycled from commercial use to amenity. Either way, cities are often linked to the name of their main river and inseparable from it. London and the Thames, Paris and the Seine, New York and the Hudson are just three examples where the river is almost as well known as the city itself.

Surabaya no longer enjoys international recognition as a river city, despite its Golden River (Kali Mas). Nevertheless, its setting for the December 2017 River Cities Symposium made it an obvious case study in how a specific riverine geography gave birth to a city and how it has survived and prospered through an often turbulent history. In the modern era, technology in the form of water control, railways, and roads has freed Surabaya of most of its riverine constraints. Nevertheless, the natural environment is never vanquished. As the city's built environment has expanded its footprint of hard surfaces, the city has become more vulnerable to monsoonal flooding exacerbated by climate change.

The second part of this chapter broadens the perspective by considering how Surabaya compares with other river cities in Indonesia. The third part focuses on the issue of scale and how the dramatic increase in the size of these cities has been at the expense of the natural environment. The final part will focus on the issue of sustainability. Climate change is increasing the vulnerability of river cities through rising sea levels and increased rainfall. If in the past the river was a solution, increasingly it will become a predicament. What lessons can be learned from how "river cities" have adapted to past tribulations and what are the best policy choices for the future?

Surabaya: A Short Story

Nature does not bestow its gifts evenly.[1] By a quirk of geography, Java's only two long rivers, the 600-kilometer Solo and the 320-kilometer Brantas, almost

1 Except where otherwise cited, this section draws on Dick (2002). Detailed maps and exposition can be found therein.

Figure 2.1 Map of Eastern Java. Navigable rivers and main towns, ca. 1900 (Adapted by Howard Dick, based on *Officieele Reisgids, 2 november 1939–1 mei 1940*)

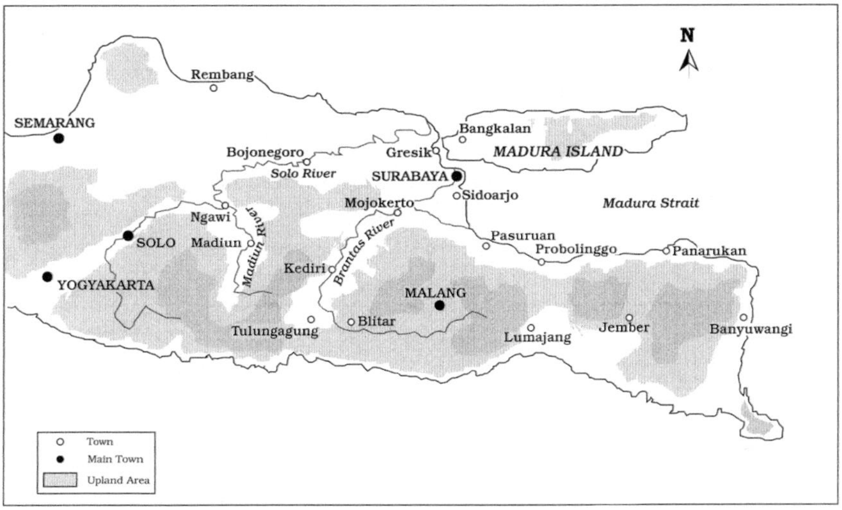

converge at the western end of Madura Strait (see Figure 2.1). The Solo River (Bengawan Solo) drains a catchment extending to the old royal capital of Surakarta (Solo) in Central Java and, by its tributary, to the hinterland of Madiun, before following a course through the northern limestone hills to Madura Strait.

The Brantas follows a half arc from near the south coast around East Java's central volcanic massif through Kediri and thereby drains much of the most fertile agricultural land in East Java. Because the western entrance to Madura Strait is protected in the lee of the island of Madura, the two rivers gave rise at their mouths to adjacent ports barely 20 kilometers apart, namely Gresik on the Solo River and Surabaya on the Brantas.

German resident and historian of Surabaya G.H. von Faber (1953) speculatively mapped the topography of Surabaya of about a thousand years ago around the time of the first known settlement. The land had not yet prograded as far as its present extent and the arm of the Brantas known as the Kali Mas (Golden River) was a wide estuary with slightly raised islands suitable for settlement. In the absence of documentary evidence, it has to be presumed that the main activities would have been fishing, boat-building, and small-scale trading. The founding of Surabaya is dated by mainly oral sources to 1293, when an invading Mongol force seeking to enforce tribute upon the kingdom of Singhasari was out-maneuvered and forced to retreat by Prince Raden Wijaya, who then founded the kingdom of Majapahit with its capital beside the Brantas on a site now known as Trowulan near Mojokerto. As the kingdom flourished, Surabaya seems to have become a control point for the main port further upstream near the royal capital (Faber 1953).

In the sixteenth century, after the eclipse of Majapahit, Surabaya emerged in its own right as one of the more prominent port city kingdoms along the north coast (*pasisir*) of Java along with Banten and Demak. Underpinning this power shift was the rise of Islam. Early in the fourteenth century the ruler of Melaka, the main entrepôt of Southeast Asia, converted to Islam, which respected trade and commerce and through trade spread to the ports along the north coast of Java. The grave of Sunan Ampel (d. 1481) in North Surabaya is the site of the city's oldest mosque. Surabaya's prosperity was maintained and perhaps enhanced after the Portuguese captured the entrepôt of Melaka in 1511.

Surabaya's independence was lost not to the Portuguese or the Dutch but to Sultan Agung, ruler of the Central Javanese kingdom of Mataram, who sought to subdue the north coast kingdoms and direct their trade though his own ports of Jepara and Tuban. Surabaya held out much longer than Gresik but after five years of siege it was subdued in 1625, reputedly after the water supply was blocked. Its population had been estimated by Dutch observers to have been 50–60,000 (Reid 1988, 17). Sultan Agung failed in his campaigns to evict the Dutch from West Java and after his death in 1646 less capable rulers were alternately compromising and at war with the United East India Company (Verenigde Oost-Indische Compagnie/VOC). The Dutch established a small settlement there in 1678 after assisting Mataram's Amangkurat II to defeat the rebel Madurese prince Trunajaya, who had made Surabaya his capital. Then in 1746, in return for Dutch assistance in Mataram's further dynastic wars, Pakubuwana II ceded to the VOC the coastal strip along the north coast of Java with all its ports, including Surabaya. The Dutch then built a modest fort beside the Kali Mas north of the Javanese town but its small factory remained subsidiary to Semarang, which was closer to (and easier to control than) the royal centers of Mataram. The depopulation of the Pasuruan/Malang district to the north and east of Surabaya during the eighteenth century wars Mataram and the VOC fought against the rebel forces of the Balinese prince Suropati and his descendants did not help (Elson 1984, 1–6). Not until the nineteenth century did Surabaya begin to re-emerge as a significant port city.

Herman Willem Daendels, the Napoleonic governor-general (1808–1811), recognized the city's strategic value and sought to strengthen its naval and military fortifications to withstand British assault. In the event, the Dutch surrendered in Central Java before these defenses were tested but after their return in 1816 more attention was paid to Surabaya. Between 1837 and 1845 the city's defenses were still being strengthened by the construction of an encircling wall and moat, a defensive technology that was already obsolete (Faber 1931).

Between 1830 and the 1930s, colonial Surabaya and its fertile agricultural hinterland became an engine for the industrial revolution in the Netherlands, exporting commodities and importing textiles and other manufactures. River and port were the vital nexus of that trade. Beginning with the system of forced cultivation in 1830 sugar became the prime export crop in the lowland valley of the Brantas,[2] which itself was the main artery to transport the product from the riverside mills to the port of Surabaya. Other exports were coffee, tobacco, and indigo.

To regulate the flow of the urban river and reduce siltation in the port, a new main course of the Brantas was cut in a straight easterly line through Porong district to discharge floodwaters into Madura Strait with a current capacity of 1500 cubic meters per second (Sisda Brantas 2018). The flow in the northern branch known as the Surabaya River (Kali Surabaya) was reduced to a current capacity of 370 cubic meters per second. Just south of Surabaya city at Wonokromo more locks restrict the flow into the urban branch of the Kali Mas to a current maximum of just 20 cubic meters per second. It thereby became a canal to the port, tidal in its lower section but with the height regulated to ensure a smooth flow of sugar-laden barges (see Figure 2.2).

The navigability provided by these massive river works became less important after 1878 as a railway network was quickly laid out along the Brantas Valley and connected through to Madiun (1882) and Surakarta (1884) in Central Java. Sugar could then be railed to the great warehouses along the Kali Mas before being loaded into lighters for transport to ships anchored in the roadstead. Then in the 1910s, after great controversy, the decision was made to build a deep water harbor to the west of the Kali Mas so that sugar and other crops could be railed onto the wharf directly alongside the ship (Dick 2002). By the time this massive project was more or less completed in the 1920s, the river had all but ceased to be a transport artery, except in its

2 The Cultivation System was a colonial scheme whereby cultivators were obliged to pay taxes in kind by delivery of nominated export crops (Elson 1994).

Figure 2.2 Kali Mas at downtown Surabaya ca. 1900 with busy river traffic (Postcard, J.M. Chs. Nijland/Howard Dick's collection)

tidal section below the business district. Production and trade still followed the topography of the river valley, but the river itself had reverted to being not much more than a natural feature (see Figure 2.3).

Until the late nineteenth century, Surabaya was still a very compact city, in fact no bigger than a modest Dutch town (Dick 2002, Maps 6.3 and 6.4). Europeans had begun to escape the cramped old town by building southwards along the main highway, which ran tangentially to the winding Kali Mas but most of the land on either side was still wet rice fields fringed by Javanese *kampungs*. The big landowners and sugar mill owners occupied what might be described as stately homes in the midst of this sea of green. However, in the 1890s a substantial tract to the south of the Resident's House at Simpang was cleared for the first garden suburb, known as the Embong-wijk. After the turn of the century, the growth of the European population and the demand for a more relaxed lifestyle led to the clearance of tens of thousands of villagers and kampung dwellers for new garden suburbs extending all the way south to the main river at Wonokromo, eastward to Gubeng, and westward to Sawahan ("Riceland") (Dick 2002). In 1906, under the new Decentralization Law, this much larger area became the Municipality (*Gemeente*) of Surabaya. By the late 1920s tree-lined boulevards and garden suburbs, electric tramways, boutique shopping, cinemas, and ice-cream parlors along with the new art-deco/*de stijl* architecture in bright whitewash gave the city a thoroughly modern veneer (Faber 1936).

Figure 2.3 Upper section of the Kali Mas, shady banks cleared of squatters but silting up and no longer navigable (Photo by Howard Dick, December 2017)

Yet for all its apparent sophistication, this newly built city had weaknesses, primarily, its vulnerability to flooding. The river locks and the diversion of floodwaters southwards through Porong protected the growing city from seasonal inundation, as did the western flood basin disgorging tidally into Madura Strait later. Nevertheless, the immediate hinterland of Surabaya was low-lying swampland that over the centuries had been improved by gravity irrigation to bring a steady supply of water to its productive wet rice/sugar fields. When this land was reclaimed for garden estates with hard roadways and drains, the problem became one of how to remove excess water in the rainy season. The Dutch engineering solution was to fit large pumps that discharged into holding basins and from there into the Kali Mas and Surabaya River (see Figure 2.4). However, these measures were not fully effective in the *kampungs*, where stagnant water combined with the lack of a sewerage system became a health problem (malaria). Increasing municipal intervention in the management of the nominally autonomous *kampungs* never properly resolved the tension between the skeletal infrastructure of a modern city and the sprawling, village-like but increasingly crowded environment of the *kampungs* where most people lived.

Figure 2.4 Massive pumps discharge excess water from low-lying districts into the Kali Mas, providing a good spot for Sunday morning fishing (Photo by Howard Dick, December 2017)

Since the mid-twentieth century, road surfacing and widening have brought about a steady encroachment of hard surfaces upon soft. Until the early twentieth century, roads were no more than compacted dirt, dusty in the dry season and muddy in the wet. After bitumen became available as residue from the Wonokromo oil refinery, hard paving created all-weather surfaces that could be edged with concrete drains. Brick, cement render, and clay roof tiles (mandated to reduce fire hazard) became typical building materials. The phenomenon of urban heat islands has been well remarked on but an associated aspect is that the increasing area of hard surfaces generated run-off as a greater excess water load for residual soft surfaces. When pumps were not well maintained or became clogged with rubbish, not least plastic, local flooding occurred and could quickly bring traffic to a standstill. This matters all the more in Surabaya because the city no longer has an urban tramway system. All mobility is by roadway and as the city sprawls, population grows, and car ownership increases, traffic becomes ever more susceptible to flooding with consequent bottlenecks and gridlock (see Figure 2.5).

Over the next few decades, global warming will increase Surabaya's vulnerability to inundation from a combination of a rising sea level, urban

Figure 2.5 Maquette of Surabaya today. The Kali Mas flows from bottom-left northwards through the central spine to the seaport. Roads have allowed the low-rise city to sprawl eastwards and westwards (Maquette at Balai Kota, Photo by Howard Dick)

subsidence, heavier monsoonal rains, and more severe floodwaters from the Brantas catchment. Surabaya is better engineered and protected than the coastal districts of Jakarta and Semarang, which are more prone to subsidence and saltwater intrusion, but expensive flood control and mitigation measures will eventually be needed. Even so, the hard-built environment in extensive low-lying parts of the city may not be sustainable.

Surabaya in Comparative Perspective

On Indonesia's most populous island of Java, Surabaya may most readily be compared with Jakarta on the smaller and much shorter River Ciliwung (120 kilometers). Though draining a more restricted hinterland, the Ciliwung did provide a good site for the sixteenth-century port city of Sunda Kelapa, which the Dutch VOC conquered and occupied in 1619 (Abeyasekere 1987). Batavia was then laid out as a model Dutch town on a grid pattern of canals with a strong fortress at the river mouth to protect the port and its rich warehouses. Unfortunately, as Leonard Blussé (1986) so well explained, the clearing of the hinterland for sugar cultivation caused the river and the urban canals to silt up, and the fetid, brackish water to become a breeding place for malarial mosquitoes. By the late eighteenth century Batavia had become notorious around the world as a death trap. In the course of the nineteenth century the administrative center was moved to higher ground and eventually to the hills at Bogor (Buitenzorg), while in 1886 a deepwater port was opened a short distance east along the coast at Tanjung Priok. Since then Jakarta, as it became known after Independence, has been, like modern Surabaya, a river city by default, thereby more prone to flooding but otherwise having a river with minimal functional use.

Table 2.1 Indonesia: Population of Jakarta and Main River Cities according to Municipal Boundaries, 1930, 1971, and 2015

City	1930	1971	2015
Jakarta	533,000	4,085,000*	10,177,924*
Surabaya	342,000	1,309,000	2,848,583
Palembang	108,000	504,000	1,580,583
Banjarmasin	66,000	224,000	675,440
Pontianak	45,000	182,000	607,618
Jambi	22,000	94,000	576,067
Samarinda	11,000	91,000	855,757

Sources: Hugo et al. 1987 and BPS 2017. *Capital city/Province

Beyond Java, Indonesia's preeminent river city is Palembang, upstream on the River Musi where it approaches higher ground. Its history traces back to the kingdom of Sriwijaya (seventh–thirteenth century), whose remains are now located within a municipality of around 1.8 million people as of 2015. When established, the city of Sriwijaya with its many temples would have been on or very near the coast but it now lies about 80 kilometers upstream because over the past seven centuries the shoreline is estimated to have prograded some 70 kilometers at a rate of about 100 meters per annum (Frankel ca. 1976, 4).

The Musi and its tributaries drain a wide hinterland extending back to the dividing range (Bukit Barisan) and the river continues to be a commercial artery and port that intersects with a railway network. The city now occupies both sides of the river after being spanned in 1965 by its first Ampera Bridge and a second bridge in 1994, but is still traversed by creeks and canals that connect it to the river. Unlike modern Surabaya and Jakarta,

Palembang retains the function and identity of a river city, even though only about a third of its tributary waterways survive and these are badly clogged and polluted (Savitri 2019). Smaller river cities in Sumatra include Jambi (population 576,000) further north on the Batanghari River. Kalimantan is preeminently the domain of rivers, which drain in all four directions from the central mountain ranges, not only in the Indonesian part but also to the East Malaysian states of Sarawak and Sabah and to the enclave state of Brunei. The main cities in order of population are Banjarmasin including adjacent upstream Banjarmasin Baru on the Martapura River, a tributary of the Barito in South Kalimantan (900,000), Samarinda (855,000) on the Mahakam in East Kalimantan (Kutei), Pontianak (610,000) on the Kapuas River in West Kalimantan, plus Kuching (Metro 700,000) on the Sarawak River in East Malaysia (BPS 2017). Brunei on the river of the same name has more the character of a town, notwithstanding the royal palace and grand mosque.

Historically these river cities in Sumatra and Kalimantan prospered through trade with the upland/highland peoples in a symbiotic *hulu-hilir* (upstream-downstream) relationship (Reid 2015; Sutherland 2007). The commercially unsophisticated upland peoples such as the Dayak in Kalimantan bartered jungle products for essentials such as salt, textiles, and metals with traders who could sell them into the regional trading system. On each river the necessity of a market center gave opportunity for "men of prowess," sometimes themselves merchants, to establish petty kingdoms to control and regulate the river trade, thereby accumulating wealth, prestige, and power. Unlike the situation on some European rivers such as the Rhine and Danube, these kingdoms were not just local tollgates but political and commercial monopolies of the entire catchment. As on the east coast of Kalimantan, these were often still petty kingdoms but Banjarmasin in South Kalimantan and Brunei in North Kalimantan became powerful enough to annex adjoining rivers while Sriwijaya/Palembang grew into a loose empire.

Conversely, a powerful external empire such as Majapahit (1293–ca. 1500) could subdue rivals such as Sriwijaya and for a time gain vassalage over river kingdoms such as Banjarmasin. External relations were always in a state of flux but the necessity of power centers on each river would remain as long as there was an international demand for their commodities. And as long as rulers could guarantee security, foreign traders and merchants would come, whether seasonally or to settle in protected quarters and marry local wives. With the coming of Islam, many of these traders came to be recognized as "Malay" by their lingua franca, but there were also Chinese, mainly from the coastal province of Fujian but including Teochew

Figure 2.6 Banjarmasin on stilts, creek-side living (Photo by Howard Dick, February 2016)

Figure 2.7 Banjarmasin, roads and bridges superimposed (Photo by Howard Dick, February 2016)

in Pontianak and Hakka miners and farmers in the Sambas district of West Kalimantan (Somers-Heidhues 2003).

In later centuries Europeans brought a new element. By the eighteenth century the Dutch VOC and British East India Company were setting up small factories in some river towns such as Banjarmasin with indifferent success. In the early nineteenth century European adventurers likewise tried their hand, though only James Brooke in Sarawak would remain. The attraction of these rivers increased with the discovery of coal and later, oil (Lindblad 1988). From the late eighteenth century, a colonial presence was established, beginning with Pontianak in 1779, Banjarmasin in 1787, and Palembang in 1821. but even at the beginning of the twentieth century the colonial presence barely extended upriver, the sultans and rajas still had influence, and most commercial activity was in the hands of Chinese merchants. In Indonesia, after Independence, these kingdoms firmly integrated into a unitary state that would be oriented towards Java rather than Singapore. In 1963 Sarawak and Sabah were brought into the federation of Malaysia, leaving only oil-rich Brunei as a much truncated but still independent enclave, the sole survivor of the river kingdoms, large and small.

Within the nation-states of Indonesia and Malaysia, modernization has involved further compromises that have stretched formerly serpentine urban forms in new directions. Air travel has become the primary means of inter-island passenger movement, so that airports have had to be built at some distance on dry ground and linked by all-weather roads. Larger ships, both bulk carriers (timber, minerals, oil) and container ships have necessitated new ports being built at or beyond river mouths to allow deeper draft, which in turn requires roads for heavy vehicles. Thus, a road network has had to be superimposed by way of reclamation, and expensive bridges constructed over remaining waterways (see Figures 2.6, 2.7). In the case of Banjarmasin, for example, decisions have been made as the old town has become too cramped and expansion and new development should spill over into the adjacent city of Banjar Baru on dry ground closer to the airport. All these cities face challenges of how best to continue to compromise with their riverine environment, whether to suppress it or, more drastically, to relocate the city to higher ground.

Scale and Technology: Environmental Immiserization

These cases help to focus attention on the issue of scale (Table 2.2). The driving force behind the accelerating urban transformation since the mid-twentieth century has been the steady growth in population. Until the early nineteenth century, the population of Java had probably never been more than 5 million (Reid 1988, 13–15). Now it is almost 150 million with over half living in urban areas. Surabaya's population had been around 50,000 in the early seventeenth century, increasing to no more than 150,000 by 1905, when it might still be described as a "traditional" urban environment. Now the municipal population is close to 3 million and the metropolitan area of Greater Surabaya about double that and thus equivalent to the entire population of Java around 1830. Most of the city's population growth has been accommodated over the 70 years since Independence.

Table 2.2 Population estimates of Java and Surabaya, 1600–2015

Year	Java (million)	Surabaya
1600	5	50,000
1830	6	N/A
1905	30	150,000
2015	145	6 million*

Sources: Elson 1994; BPS 2017; Dick 2002. *Greater Surabaya

The river cities of Sumatra and Kalimantan have remained more modest in size (Table 2.1). As of 1930, only Palembang slightly exceeded 100,000 (Hugo et al. 1987, 52). These low figures reflected the overall low rate of urbanization in the colonial era—in 1930 only 6.7 percent—and also the limited populations and resources of the hinterlands of these river cities (Hugo et al. 1987, 89). The small local market and relatively high labor costs meant little scope for urban activity. High dependence on commodities also meant an uncomfortable cycle of boom and bust. Banjarmasin is a good example. In the latter part of the nineteenth century traditional reliance on jungle products and pepper gave way to *gutta percha* (jungle latex) and coal, then in the early decades of the twentieth century to rubber, in the 1960s and 1970s to timber, in the 1980s and 1990s to rattan for furniture making, then in recent years, coal in massive quantities, all without diminishing heavy reliance on a single staple and with increasing degradation of the environment.

So why have some river cities grown into cities of millions of people while others have not? As the case of Surabaya suggests, it makes a big difference whether the city has a fertile and highly productive hinterland that can itself sustain a dense population. This is also the case for Yangon (4.9 million agglomeration), Saigon/Ho Chi Minh City (7.5 million agglomeration), and Bangkok (9.4 million metropolitan only), the latter two approaching megacity status of 10 million people or more (United Nations 2016).

That such riverine cities have been able to absorb so many rural-urban migrants while still enabling their huge populations to improve their standard of living is a truly remarkable developmental feat. However, the rate of urban growth and its increasing scale and pace has involved an accumulation of environmental externalities that impair the quality of urban life and threaten the long-term viability of those built environments (Rimmer and Dick 2009).

Urban dwellers complain endlessly about the lack of infrastructure and increasing traffic congestion. There are also the worsening problems of air and water pollution. Better infrastructure may alleviate some of these problems but it will not resolve them because modern cities are antagonistic to natural rhythms and flows. Motor vehicles, for example, provide amazing flexibility but need paved roads and parking places, which have been at the expense of canals and soft surfaces. Multi-story buildings allow greater density but cannot readily be built on stilts. Buildings of concrete and glass and hard paved roads create heat islands. Air-conditioning cools the internal environment through massive consumption of energy, but adds to external heat and pollution.

The problem may be better understood from the perspective of biological systems. Any organism that expands without limit will eventually reach a threshold at which it ceases to be environmentally sustainable, with potentially catastrophic results. This is not a crudely Malthusian argument. Technological advance and human ingenuity have allowed cities to accommodate millions of people within a few decades and this population trend will persist. However, coastal and riverine cities will not be able to sustain rising living standards unless the urban environment adapts to a harsher natural environment with more sophistication than present-day concrete, high-energy jungles.

Conclusion: Living in the Anthropocene

Historically the Malay world was a water world. People lived beside the seas and rivers and accepted the tides and monsoons as part of the rhythm of life. The built environment and means of transport adapted to these rhythms and were sustainable in all but the most extreme circumstances. Houses were built on stilts beside waterways and every household had one or more boats, from simple dug-out canoes to more sophisticated *sampans* and *prahus* for local transport or trade with the wider world. Rivers provided water and sewerage for households as well as water for crops and a supply of fish for protein. Such a way of life may still be observed along the waterways of Kalimantan and Sumatra, as well as further afield along the Mekong, Irrawaddy, or Chao Phraya. With modern improvements such as outboard motors for boats, electricity for houses, and satellites for telecommunications, such lifestyles may still appear resilient.

However, there is mounting evidence that there is no longer a sustainable balance between an increasingly brutalist urban environment and a desecrated natural one. This is hardly a new dilemma. Unintended consequences always flow from developing urban hinterlands and engineering urban habitats. A classic case, mentioned above, is how in the eighteenth century the model Dutch city of Batavia became afflicted by endemic malaria when its artificial waterways and canals became clogged and fetid. A fascinating European case is how the canalization of the Rhine to free land for agriculture and make the river safe for navigation led to greater vulnerability of adjacent cities to catastrophic flooding (Blackbourn 2006).

Since the mid-twentieth century, Asia's cities have become caught in a tightening three-way vice. First, population growth, development, and deforestation in the hinterland has increased water run-off and downstream siltation. Second, rapid urbanization combined with modernization in the form of paved roads and brick and concrete buildings has created heat islands, reduced the absorption of urban surfaces, and in some cities led to irreversible subsidence. Third, climate change is leading to an irreversible trend of rising sea levels while increasing the intensity of storms. Estimates of sea level rise vary according to assumptions about policy action and the melt rate of the Arctic and Antarctic icecaps but the scientifically agreed range is now between 0.5 and 1.0 meter by 2100 (AMAP 2017). Periodic, partial, and increasingly severe inundation is inevitable for cities which are at or near sea level. If they can no longer function as their sophisticated networks and technologies require, they will atrophy.

Nature and science may be ignored except for a short-term narrative of disaster management, but stupidity is not a strategy for survival. Most immediately, river cities face the challenge of balance with the water world. More investment in the clumsy, environmentally damaging hard infrastructure of roads, parking, and energy-inefficient buildings is not a solution. Twentieth-century technologies of consumption and investment are not sustainable without environmental adaptation. Fortunately, some of those technologies are becoming available: renewable energy can substitute for fossil fuels in generating power; electric- and hydrogen-powered vehicles can reduce the use of polluting petrol and diesel; building codes can mandate energy-efficient buildings; closed-loop recycling can eliminate mountains of urban waste and sewerage. All these things are do-able if there is enough public pressure to expose the follies of environmental desecration and to shame the vested interests that perpetrate them and the politicians who dance to their tune. All these things are matters for local governments and every city has a role to play. It is not only a challenge of how to build and where to build, as of how to live. Networks of cities will invent the future, not meetings of national leaders.

Because a two percent rise in global temperature is already "locked in," there is also a need to adapt to the climate change that will occur. As the historical experience of the Netherlands shows, it is possible to live below sea level if there is enough investment in and maintenance of dykes, drainage, and pumps. London and Venice have made huge investments in barrages to protect themselves against tidal and storm surge. Southeast Asia's river cities may adopt a similar strategy but the cost is far beyond anything that municipal budgets can afford. In the case of Indonesia, the cost of such public works for Jakarta, Surabaya, Semarang, Palembang, Pontianak, and Banjarmasin, among other river cities, is probably beyond the foreseeable capacity of national budgets, even with a substantial rise in taxation. Zoning and building codes that accommodate flooding and natural forms of flood mitigation may buy time.

Nevertheless, some river cities may have no realistic alternative but to relocate to higher ground. This might seem an absurdly radical and costly measure but in fact Southeast Asia's royal capitals have always been transient, albeit some more so than others as over the past millennium in Java, Thailand, and Myanmar (Reid 1988). War and defensive needs were often the cause and motivation, but politics and environmental issues also played a role. From the fifteenth century the locus of Thai power moved downstream from Sukhothai to Ayutthaya, then after the Burmese conquest of 1767 further downstream to Thonburi/Bangkok. In Burma the movement was in the opposite direction, from Pegu upstream to Ava/Mandalay (Reid 2015). In 1746 the humiliated sultan Pakubuwono II moved the court of Mataram from the battered city of Kartasura to its nearby and present site of Surakarta (Ricklefs 2001). In modern times, Malaysia's prime minister Mahathir moved the seat of government from Kuala Lumpur to nearby Putra Jaya in 1999 (King 2008), then in 2006 Myanmar's military rulers moved the capital from colonial Rangoon (Yangon) inland to Naypyidaw, some 360 kilometers to the north. Putra

Jaya and Naypyidaw are aspirational cities, being both modern and model capitals in a tradition that blends Asian, Middle Eastern, and Western concepts and styles. Naypyidaw marks a historic retreat from the colonial capital of Rangoon, which in 1886 officially succeeded the former Burmese capital upstream at Ava/Mandalay. Bangkok, like its predecessor Ayutthaya, remains defiantly downstream while inexorably sinking. It has no nearby higher ground to which to retreat. Pre-Ayutthayan capitals like Sukhothai were far away in upland Thailand.

Jakarta has been relocating ever since 1810, when Governor-General Daendels moved the seat of government from Kota to Weltevreden-Harmoni around what is now the presidential palace and the Department of Finance. In the 1940s the returning Dutch founded a garden suburb at Kebayoran Baru. After Independence, President Sukarno, architect and urban planner, was instrumental in developing a new city center around Jalan Thamrin. Surabaya, likewise, moved south towards Wonokromo and beyond as well as to the east and west. Banjarmasin is expanding upstream around Banjar Baru. Climate change is slow enough that relocation can still be done in an orderly way. In the case of Jakarta, the government of President Joko Widodo has decide to gradually relocate the capital to East Kalimantan. In the case of Surabaya, the retreat may need to go as far as the higher ground to the south of the Porong River.

The lesson of climate change is that, in the very long-run, everything changes. Cities are being renewed all the time to increase density, repair damage, or meet the demands of economic and social restructuring. There is no reason why renewal should occur in the same place. In some ways it is easier and less disruptive if it does not. In a familiar "doughnut" pattern, new industries and housing estates take up "greenfield" sites, usually agricultural land, beyond the city limits. So the issue is therefore not whether cities should expand but how far and in which direction(s). Coastal and riverine cities will bear the initial brunt of climate change so they have the greatest need to innovate. It is not smart just to go on repairing the annual flood damage. Municipal budgets may presently lack the funds to do more, but wealth enough is concentrated in these cities. The political issue is how those funds can be mobilized and whether governments can begin to work with communities to articulate and share a common interest for future generations. Nature will not forever be conquered. Children understand that instinctively because they still have a sense of wonder not yet dulled by ambition and greed. Responsible adults should not be more frightened to confront reality.

References

Abeyasekere, Susan. 1987. *Jakarta: A Short History*. Singapore: Oxford University Press.

AMAP. 2017. "Snow, Water, Ice and Permafrost. Summary for Policy-Makers." Arctic Monitoring and Assessment Programme, Oslo. Available at www.amap.no/swipa2017, accessed 23 April 2018.

Blackbourn, David. 2006. *The Conquest of Nature: Water, Landscape, and the Making of Modern Germany*. New York, NY: Norton.

Blussé, Leonard. 1986. *Strange Company: Chinese Settlers, Mestizo Women and the Dutch in VOC Batavia*. Dordrecht: Foris.

BPS. 2017. "Penduduk tengah tahun 2011–2015 [Inter-censal population], hasil proyeksi," data table.

Dick, Howard. 2002. *Surabaya, City of Work: A Twentieth Century Socioeconomic History*. Athens, OH: Ohio University Press/Singapore: NUS Press.

Elson, Robert. 1984. *Javanese Peasants and the Colonial Sugar Industry: Impact and Change in an East Javanese Residency, 1830–1940*. Sydney: Allen & Unwin.

Elson, Robert. 1994. *Village Java under the Cultivation System, 1830–1870*. Sydney: Allen & Unwin.

Faber, G.H. von. 1931. *Oud Soerabaia: De Geschiedenis van Indië's Eerste Koopstad van de Oudste Tijden tot de instelling van de Gemeenteraad*. Surabaya: Gemeente Soerabaia.

Faber, G.H. von. 1936. *Nieuw Soerabaia: De Geschiedenis van Indië's Voornaamste Koopstad in de Eerste Kwaarteeuw sedert Hare Instelling, 1906–1931*. Surabaya: Van Ingen.

Faber, G.H. von. 1953. *Er Werd een Stad Geboren: De Wordgeschiedenis van het Oudste Soerabaja*. Surabaya: Kolff.

Frankel, Ernst G. ca. 1976. "The Palembang Port and Shipping Study," VII: I (USAID). Available at http://pdf.usaid.gov/pdf_docs/PNAAN194.pdf, accessed 30 April 2018.

Hugo, Graeme, Terry Hull, Valerie Hull, and Gavin Jones. 1987. *The Demographic Dimension in Indonesian Development*. Singapore: Oxford University Press.

King, Ross. 2008. *Kuala Lumpur and Putra Jaya: Negotiating Urban Space in Malaysia*. Singapore: National University of Singapore Press.

Lindblad, J. Thomas. 1988. *Between Dayak and Dutch: The Economic History of Southeast Kalimantan*. Dordrecht: Foris.

Reid, Anthony. 1988. *Southeast Asia in the Age of Commerce, 1450–1680: The Lands below the Winds*. New Haven, CT: Yale University Press.

Reid, Anthony. 2015. *A History of Southeast Asia: Critical Crossroads*. Oxford: Wiley Blackwell.

Ricklefs, Merle. 2001. *A History of Modern Indonesia since c.1200*, 3rd edn. Houndmills, UK: Palgrave.

Rimmer, Peter, and Howard Dick. 2009. *The City in Southeast Asia: Patterns, Processes and Policy.* Singapore: National University of Singapore Press.

Savitri, Yulia. 2019. "Musi River's Tributaries Diminishing." *The Jakarta Post*, 23 March. Available at https://www.thejakartapost.com/news/2019/03/23/musi-rivers-tributaries-diminishing.html, accessed 30 April 2018.

Sisda BBWS Brantas. 2018. *Profil BBWS Brantas*. Surabaya: Balai Besar Wilayah Sungai Brantas.

Somers-Heidhues, Mary. 2003. *Golddiggers, Farmers and Traders in the "Chinese Districts" of West Kalimantan, Indonesia*. Ithaca, NY: Cornell University Press, Southeast Asia Program.

Sutherland, Heather. 2007. "Geography as Destiny? The Role of Water in Southeast Asian History." In *A World of Water: Rain, Rivers and Seas in Southeast Asian Histories*, edited by Peter Boomgaard, 27–70. Leiden: Brill.

United Nations. 2016. *The World's Cities in 2016*. Data Booklet. Available at: http://www.un.org/en/development/desa/population/publications/pdf/urbanization/the_worlds_cities_in_2016_data_booklet.pdf, accessed 20 November 2017.

Biography

Howard Dick is a professorial fellow in the Faculty of Business and Economics at the University of Melbourne and writes in Economic History and Urban Studies with prime focus on Indonesia. Books include *Surabaya, City of Work* (2002) and, as co-author, *Cities, Transport and Communications: The Integration of Southeast Asia* (2003) and *The City in Southeast Asia* (2009).

3 From the City to the Sea

Riverside Temple Networks in South India[1]

Emma Natalya Stein

Abstract

The riverside city of Kanchi, South India, was a major commercial center, royal capital, and holy destination during the eighth through thirteenth centuries. Praised as a jewel in early Tamil literature, Kanchi's Palar River, and its tributary Vegavathi, provided the lifeblood of the city. However, today the Palar is dry and the Vegavathi is a polluted drainage route. Consequently, the settlements in Kanchi's vast suburban expanse have become remote, disconnected villages—a network of stone temples is all that survives as testament to each place's former prosperity. This chapter argues that the rivers contributed to Kanchi's urbanization process in the premodern period, and that the rivers' deterioration in recent centuries has contributed to the disappearance of places that previously thrived.

Keywords: urbanism, South Asia, Southeast Asia, temples, architecture, Hinduism

1 The research for this paper is drawn from my doctoral thesis, *All Streets Lead to Temples: Mapping Monumental Histories in Kanchipuram, c. 8th–12th Centuries CE* (Yale 2017), and book, *Constructing Kanchi: City of Infinite Temples* (Amsterdam University Press 2021). The book offers a first-ever understanding of Kanchi's transformation from a relatively small settlement into a major metropolis, popularly known as the "city of one thousand temples." Fieldwork was carried out during 2013–2016, thanks to generous support from the American Institute of Indian Studies (AIIS), the Smithsonian, and the Yale South Asian Studies Council and Department of the History of Art. The IIAS/UKNA symposium (2017) gave me the opportunity to further develop my research on historical connections between rivers and cities in northern Tamil Nadu. Research into the more recent disappearance of rivers was carried out during a landscape workshop at the City Palace Museum, Udaipur, Rajasthan, 6–7 March 2019, with sponsorship from the Smithsonian and the Maharana of Mewar Charitable Foundation.

Padawangi, Rita, Paul Rabé and Adrian Perkasa, *River Cities in Asia*.
Taylor & Francis Group, 2022
DOI: 10.5117/9789463721851_CH03

Figure 3.1 Veṅkatīśvara (Vishnu) temple, Tirumukkūṭal, Tamil Nadu, South India, temple construction begun ca. ninth century, seen from the dry riverbed of the Palar, Ceyyar, and Vegavathi confluence (Photo by Emma Natalya Stein)

In the village of Tirumukkūṭal, in Tamil Nadu, South India, a stone temple dedicated to the Hindu god Vishnu stands at the confluence of three rivers. Situated on the southern bank, the temple faces north to overlook the rivers, as if the waters were to bathe the god's feet. Despite the centuries of renovations and reconstructions that this temple has undergone, Vishnu's northward gaze has remained steady. When new enclosures were added to the complex over time, the architects included a screened window in each wall on the northern side, so that the god's view would not be interrupted. However, the rivers Vishnu oversaw did not survive. Today, the three rivers that once swirled at his feet are nothing more than dry beds of sand (Figure 3.1).

While the Vishnu temple at Tirumukkūṭal occupies a particularly dramatic location, it is by no means the only riverside shrine in its vicinity. The village is situated 20 kilometers to the east of Kanchipuram, a major South Indian city that held elite status as a royal capital, commercial crossroads, and devotional epicenter in the eighth through thirteenth centuries (Stein 2017 and 2021). Kanchipuram—or Kanchi as it is more simply known—also stands on the banks of a river and is surrounded by temple sites that track along historical rivers. Farther south, the Kaveri River is well known throughout India as a Tamil river, and Chennai is well

Figure 3.2 Temple sites around the city of Kanchi (Map by Emma Natalya Stein and Daniel Cole, Smithsonian Institution)

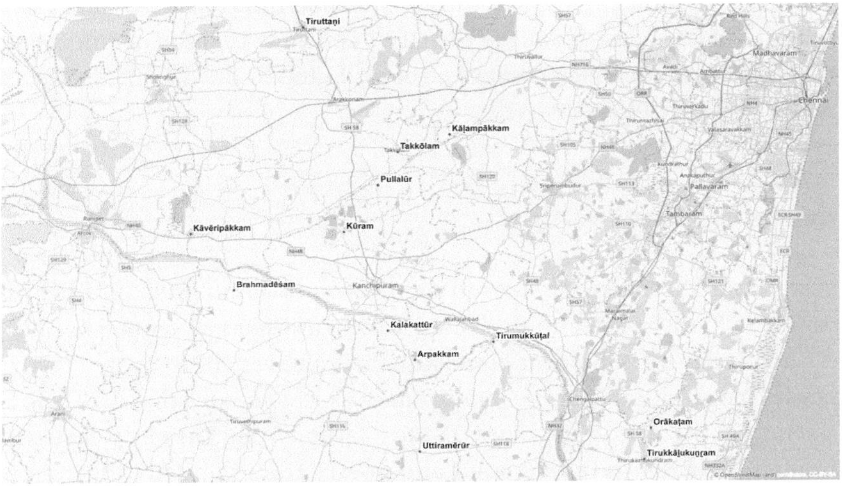

known today as a modern river city, yet rivers throughout northern Tamil Nadu were important arteries of travel and irrigation.[2] Located to the north of the Palar River, the city of Kanchi spans the Vegavathi, which branches from the Palar, arcs through the city, and then merges with the Palar and also the Ceyyar river at the triple-confluence in Tirumukkūṭal.

Tirumukkūṭal is now a mostly destitute village. Its single, stark temple with little ritual life contrasts wildly with the vibrant exuberance found at many temples inside the city. Kanchi is in fact surrounded by places just like this—settlements housing stone temples that were built up incrementally over the course of the city's 500-year urban efflorescence. Today most of these settlements constitute little more than remote villages lying well outside circuits of travel and exchange. Their temples are the only vestiges of a much more colorful past. By mapping the locations of these monuments, I have found that the early settlements around Kanchi mark out the pathways where people once traveled, and rivers once flowed. Today holding no water, some of these riverbeds are scarcely visible from ground level. However, aided by satellite imagery, the temple sites reveal that northern Tamil Nadu was once a land flowing with rivers (Figure 3.2).

In this chapter I propose that the city of Kanchi incrementally developed outwards from its center during the course of the eighth through thirteenth centuries, and more specifically that rivers determined the vectors of this

2 For discussion of water's edge urbanism in Chennai, see Chapter 9 in this volume.

expansion. In order to trace the ancient river routes, I begin by situating Kanchi as the major royal capital, commercial center, and multireligious pilgrimage destination that it was in the premodern era (Stein 2017 and 2021). I then explore key sites outside the city. I conclude that the deterioration of the rivers in recent centuries has contributed to the disappearance of places that previously thrived.

Kanchi as a Temple-City

When the Pallava dynasty moved southwards to Tamil Nadu in the third century CE, the royal family selected Kanchi as their new capital because it was already an important place. Kanchi is an even more ancient city. It houses megalithic burial sites and archaeological remains that date before the turn of the first millennium (Foote 1865 and 1916). The city is featured frequently in South Indian epics and poetry from the Caṅkam era (ca. first century BCE–sixth century CE). One text praises the city as a place populated by foreign merchants and resounding with diverse festivals (*Pattuppāṭṭu* 2012, 184). Kanchi is described as a major center of Buddhist learning in the epic *Maṇimēkalai* (Shattan 1993). This reputation is further maintained in records from the wider Buddhist world. When the Chinese pilgrim Xuanzang visited Kanchi in the middle of the seventh century, he praised its many shrines and the multitude of monks that tended them (Xuanxang 1884, 228–229).

Kanchi continued to maintain a position of importance within literature of the late first and early second millennium CE. A great many of Kanchi's temples are praised in the poetic verses of the Tamil saints.[3] The life stories of these poets were later compiled into a hagiographic anthology, called the *Periya Purāṇam*, which was composed in the twelfth century at the Chola court.[4] All of these texts extol Kanchi's temples, its cosmopolitanism, and its plentiful natural resources.

Kanchi became fully realized as a courtly urban hub between the eighth and thirteenth centuries (Stein 2017 and 2021). In that pivotal 500-year period, the city served as the royal capital for two major dynasties, the Pallavas and then the Cholas, and it was home to a large and diverse population. The Pallava rulers fashioned themselves as great patrons of arts and literature

3 The *Tēvāram* and *Divya Prabandam* are anthologies of hymns composed by the Śaiva *nāyaṉmārs* and Vaiṣṇava *āḻvārs* respectively.

4 Versions consulted: Cēkkiḻār (1950, 1990, 1995, 2006).

in part by building temples.[5] They selected stone, a durable and costly material that they dubbed the hallmark material of South India's elite. The Pallava's sacred monuments were carved from fine-grained sandstone, their surfaces painted in jewel-toned pigments. They were crowned with pyramidal towers and elegantly ornamented with figures of deities and inscriptions extolling the royal family.[6] When the Cholas supplanted the Pallavas in the middle of the ninth century, systems of temple patronage also changed.[7] Whereas most Pallava temples had been royal establishments, local communities were the sponsors of most Chola-period temples. To maintain these establishments, the Chola kings made gifts of tax-free land to agrarian and religious communities throughout their expanding domain.

Famed for its manifold temples as well as its commerce, Kanchi continually drew attention from poets, merchants, entrepreneurs, Hindu devotees, and pilgrims from across the Buddhist world.[8] Perhaps because of this diverse appeal, the city took shape with a somewhat unconventional urban form. Kanchi can be described as a "temple-town," in that every aspect of the city's schedule revolves around the temples. In addition to the local residents who make regular visits, busloads of devotees on extended pilgrimage tours flood the city daily.[9]

Much as the temples dominate the urban ecology, however, an important distinction exists between Kanchi and other so-called "temple-towns" in Tamil Nadu. Such settlements are typically clustered around a single, expansive temple (Michell 1993). For example, at places like Tiruvaṇṇāmalai, the main temple stands in the middle of the settled area (Figure 3.3). The primary sanctum lies at the temple's center, surrounded by enclosure walls punctured in all four cardinal directions by gates with soaring towers. A sequence of streets is laid in concentric squares parallel to the outer walls of the sacred complex. Perpendicular streets lead through the town directly to the gateways. These temple-complexes were typically built up over the course of multiple centuries. Often, the shrine at the center of the temple is the complex's oldest part. Over time, additional shrines were built and

5 For discussion of the Pallavas, see Gillet (2010).

6 For Kanchi's dynastic history, see Mahalingam (1969); C.R. Srinivasan (1979); and K.R. Srinivasan (1983a and 1983b).

7 Chola patronage is currently under reconsideration by scholars. Orr (2006); Kaimal (1996); Francis (2014); Dehejia (2016).

8 As late as the fourteenth century, Kanchi was renowned in the East Javanese *Deśavarṇana* (*Nagarakṛtāgama*). Prapañca (1995). For discussion, see Stein (2017, 306).

9 A temple like Ekāmbaranātha receives observably close to a thousand visitors on a daily basis.

Figure 3.3 Typical "temple-town" with the Aruṇācalēśvara (Shiva) temple at its center, surrounded by the settlement of Tiruvaṇṇāmalai, Tamil Nadu, South India, temple construction begun ca. seventh century (Photo by Emma Natalya Stein)

then enclosed with new walls so that they too became parts of the temple. The walls and their gateways get taller—and more recent—as the distance from the center increases.

Kanchi's layout is strikingly different (Figure 3.4). There is no single, central monument. Rather, in Kanchi literally hundreds of temples pepper the full urban expanse, and the city squeezes between them. Vestiges of the Pallava- and Chola-era temples can still be found, either as independent structures or parts of later buildings. Fragments are often built into walls, reused as paving stones, painted over, or otherwise disguised or transformed. Until my research, these remains had never been systematically documented or mapped. As I recorded through extensive ground surveys and archival research, at least eight surviving full temples were established in the eighth century alone, but dozens more small shrines and scattered fragments serve as evidence of additional Pallava-era temples. No less than 25 extant shrines date to the Chola period, but disengaged sculptures tell us there were even more (Stein 2017 and 2021).

Kanchi's temples range widely in date of construction and also devotional traffic. While a handful of temples have been preserved as monuments under the official protection of the Archaeological Survey of India, far more remain magnets for devotees (Figure 3.5). The status of temples fluctuates.

Figure 3.4 Temples in Kanchi, ca. eighth to thirteenth century, showing distribution and chronology within the urban landscape (Map by Emma Natalya Stein and Daniel Cole, Smithsonian Institution)

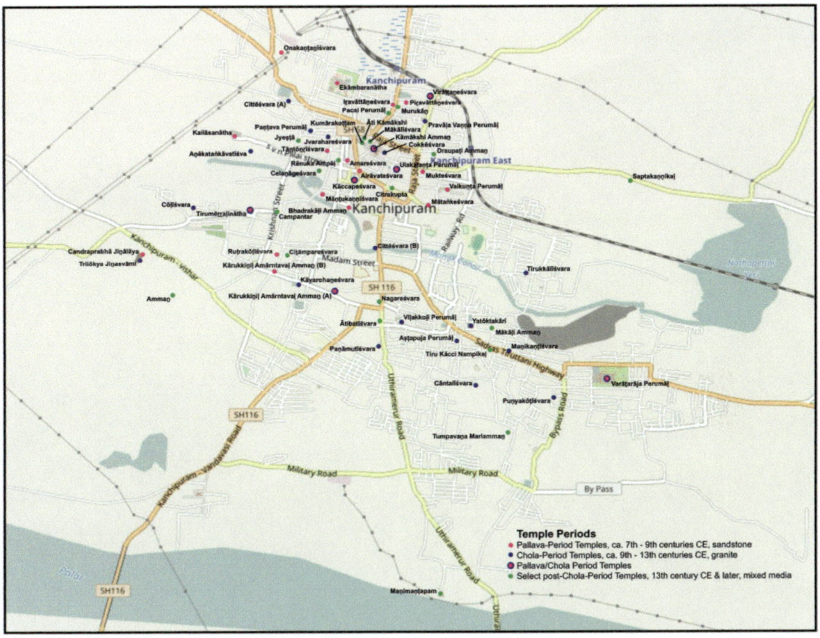

Figure 3.5 Ekāmbaranātha (Shiva) temple, Kanchi, a destination for pilgrimage tours, temple construction begun ca. sixth century (Photo by Emma Natalya Stein, 22 June 2015)

Figure 3.6 Bronze festival icon of Vishnu in procession, carried by priests from the Varadarāja Perumāḷ temple, Kanchi (Photo by Emma Natalya Stein)

At present, the Kāmākṣi Ammaṉ temple is the most prominent pilgrimage destination in the city. Dedicated to the Goddess, this large complex is typically thronged with visitors both day and night, and it has recently hosted some of the city's most elaborate ceremonies. However, scholars have argued that prior to the fourteenth century, this site was little more than a loose cluster of Buddhist and Jain shrines that had yet to be converted into a unified Hindu temple-complex (Venkataraman 1973; Dehejia 1988).

Kanchi's gods are made mobile in processions. Bronze sculptures are understood to be the real embodiments of Hindu deities, such as Shiva, Vishnu, or the Goddess. Dressed with fine textiles and adorned with jewelry made of gold studded with gemstones and pearls, portable bronze statues of the gods are regularly carried through the streets, as they were more than 1000 years ago. Because vision is a central component of ritual practice in Hinduism, bringing a divine image into public view enables the greatest number of people to receive the blessings of the gods (Figure 3.6).

Processions extend the sacred zones beyond the temple walls. Historically, the pathways they followed also functioned as a means through which to establish economic and political ties with areas beyond the city's borders (Orr 2004, 437–470). Some processions carried the gods to temples upwards of 100 kilometers away. Along the journey, they stopped intermittently at settlements where there were temples. The party would find ready hosts in the communities that maintained the temples and cultivated the surrounding lands. Consequently, each pause allowed time not only for rest and ritual, but also for conducting diplomatic relations between Kanchi and places outside the city. In the eastern direction, Kanchi was connected with an array of coastal settlements, some of which were accessed by following the course of the Palar River.

Following the Rivers

In Tamil literature and temple legends, Kanchi's Vegavathi river is often praised as a jewel of the city. Its waters are considered holy, and it is also personified as a great devotee of the gods. Both drought and flooding are persistent and ever-increasing problems for present-day Tamil Nadu, and numerous ancient myths and legends tell of the Hindu deity Shiva or Vishnu causing rivers to swell and flood, or to divert or halt their flow. In addition to the manifold problems associated with flooding in urban areas, such as Chennai, floods in rural areas have the capacity to cause entire settlements to relocate.[10] For example, the village of Kalakattūr relocated in response to a flood. Accessing the ancient, multi-temple village site now requires a 2-kilometer walk along the high bund—a raised mound of land. According to the village priest, some 300 years ago there were plans to pave this road, but the lake swelled in retaliation and broke the bund, flooding the village (pers. comm. Feb. 2016). Frightened, the residents fled and shifted the village to its present location. In the process, the village seems to have shrunk significantly, as the now-overgrown expanse of the earlier village is quite extensive.

The Vegavathi and its parent, the Palar, are typically described as seasonal rivers in the literary accounts. For example, a passage in the twelfth-century *Periya Purāṇam* reads:

> The river Pālāru is fed by milk from the udder of Surabi, the celestial cow. It flows down Mount Nandi in a torrent and fills the lotus-covered tanks, sweeping down pearls and jewels, sandalwood and eagle wood in its course. As milk flows in abundance from the mother's breast at the touch of her child, so in the hot season farmers dig in the sand dunes of the riverbed until the water gushes out. Then it flows through the channels into the low-lying fields, and as the level rises, carries away the sluices intended to control its flow. Other rivers too flow among the broad fields to fill the tanks. When the water is let out through the sluices and is distributed abroad, the farmers raise a shout of joy. (Cēkkiḻār 2006, 109–110)

Literature also often associates rivers with temples. For example:

> In this region there are many Shiva temples built on the banks of the broad rivers … The presence of all these temples is surely the fruit of the penances performed by the people of this agricultural region. (Cēkkiḻār 2006, 110)

Of Kanchi specifically, the poet-saint Campantar sings:

> In bustling Kanchi … those who give alms will not think of anything else except the Lord [Shiva] who has Ekāmbaranātha temple on the bank of the river Kampai [an older name for Vegavathi], where *koṉṟai* flowers blossom like gold. (Campantar 2.12.5)

10 For more on flooding and its implications for modern urbanism, see the essay by Kuei-Hsien Liao (Chapter 10) in this volume.

Inscriptions

While the literary references give us a sense of the importance of rivers as evocative and sacred spaces in the poetic imagination, inscriptions remind us that the metaphorical accounts were grounded in real world topographies. Inscriptions were a means through which a ruler could simultaneously establish and document his claims to a growing geographical expanse and sphere of influence.[11] Inscriptions in Tamil Nadu are typically carved into the walls of stone temples or incised into sets of copper plates. Sometimes they can also be found on loose or built-in stone slabs that survive from earlier structures.

Inscriptions provide important information, but they usually do not answer many of the questions that present-day scholars want to pose: Who built the temple? How long did it take? What were the motives behind its construction? And just what was the role of the king? Rather, inscriptions document legal determinations concerning temple property, and they record pious donations, such as supplies, lamps, and divine bronze sculptures, which were offered to the temple's gods by individuals or collectives of devotees. In this way, temples become archives of certain aspects of their own histories.

Inscriptions also work to establish boundaries that distinguished settled areas from the surrounding wilderness. The line was not specified unless another settled area lay immediately on the other side (Heitzman 1987, 799).[12] Rivers or other bodies of water sometimes offered natural boundary markers. For example, an inscription on the Shiva temple in the village of Tiruttaṇi (40 kilometers north of Kanchi) uses surrounding canals and channels as its points of reference (Gillet 2016, 478). Inscriptions could also specify which bodies of water a settlement was permitted to use. An example is found in the Kaśākūṭi copper plates, which define the borders for a village called Ekadhīramaṅgalam (Hultzch 1992, 342–361, no. 73). The inscription states that irrigation channels could be dug from the Vegavathi and Ceyyar rivers and from a nearby reservoir. In an agrarian, primarily rice-based economy, access to irrigation was of the utmost necessity.

Inscriptions also mention different types of water bodies, which shows there was a keen attention to the diverse scales and functions of both natural and man-made water resources. Construction and maintenance of irrigation networks was partly state-run and partly due to the efforts of private individuals (Minakshi 1938, 94). Types of constructed water bodies

11 For an excellent explanation of inscriptions in Tamil Nadu, see Orr (2006).

12 See also Heitzman and Rajagopal (2004).

known from Pallava and Chola inscriptions include reservoirs (*taṭaka*, *ēri*), wells (*kēṇi*), canals (*kurangu*, *kāl*, *kiḷaikkāl*, *ōḍai*), picotahs (*kuṟṟetam*, *pērēttam*), and sluices (*kalingu*) (Minakshi 1938, 94–106). Many of these can still be found in various conditions and levels of use. Water was collected in the reservoirs from three sides whose boundaries expand and contract seasonally.[13] It was then disseminated from the reservoir's remaining side to the surrounding paddy fields through a bund by means of a sluice gate that could be opened or closed. This system of water management is still widespread in Tamil Nadu. Because tanks and reservoirs historically were so important and so prominent, a special committee was convened annually to oversee their maintenance in a given municipality (Minakshi 1938, 106–107).

Throughout the Tamil landscape, contiguous settlements were described using a hierarchy of subdivisions. Together with the irrigation networks and reservoirs, these subdivisions formed constellations of places that were connected both geographically and legally. Like the water bodies, specific terminology referred to the different classifications of settlements.[14] *Tani-ūrs* (independent villages) were clustered together into *nāṭus*, which were typically named for the primary village in the group. The *nāṭus* were then grouped into *kōṭṭams*, which in turn lay within the larger category of *maṇḍala*. For example, eleventh- and twelfth-century inscriptions on Arpakkam's Shiva temple refer to Arpakkam as lying within Mākaṟal-*nāṭu*, a subdivision of Eyiṟ-*kōṭṭam*, a district of Jayaṅkoṇḍacōḻa-*maṇḍalam* (Mahalingam 1989, 110, Cg.-461–462). This means that Arpakkam was a *tani-ūr* situated within the cluster of villages in which Mākaṟal was considered the main place. Mākaṟal belonged to the consortium called Eyiṟ. And all were under the dominion of the Cholas in northern Tamil Nadu, called Jayaṅkoṇḍacōḻa-*maṇḍalam*. Whereas Cōḻa-*maṇḍalam* designated the Kaveri area hundreds of kilometers south, Jayaṅkoṇḍacōḻa-*maṇḍalam* centered on Kanchi.

Although they rarely mention construction directly, inscriptions chronicle the temple's period of greatest use. For example, most of the physical structure of the present-day Vishnu temple-complex at Tirumukkūṭal dates no earlier than the sixteenth century. However, inscriptions reveal that the temple was most active several hundred years before then. Considered a naturally occurring sacred place in India, the confluence of rivers likely

13 Coelho discusses the often "smudgy" boundaries between water and land throughout India (see Chapter 9 in this volume).

14 For longer explanations of the hierarchies of lands and the terminology used in inscriptions, see for example Burton Stein (1980); Champakalakshmi (1993); Heitzman (1997 and 2001).

held at least a humble shrine at a very early date. The first inscription on the stone structure dates to the ninth century. It is carved on a slab that has been discretely included in a more recently built wall, to the back of the sanctum. A stone temple must have replaced the perishable shrine by the time of this inscription, which tells us that the Pallava king, who ruled from his capital at Kanchi, gifted a lamp to burn perpetually in front of the temple's god (Mahalingam 1989, 332, Cg.-1299). Several stone sculptures and architectural elements in the complex also survive from this period, such as a series of pillars that supports an entry hall, which was constructed later. The lion-shaped pillar base—hallmark of the Pallava style—finds its prototype in eighth-century temples built inside of Kanchi. By the eleventh century, Tirumukkūṭal's riverside shrine had become far more than just a temple. Not only had the stone structure been expanded significantly, the temple had also begun to function as a major municipal institution. In addition to a place for devotion, the establishment provided a large hospital, a religious school, and a public feeding house (Mahalingam 1989, 332–338, Cg.-1299–1316).

Inscriptions further enable us to trace former connections with places and communities outside the temple's immediate area. Rivers and cities feature frequently within the Vishnu temple's records. A portion of an inscription stipulates that bundles of hay collected from the local residents should be used only to fund a temple flower garden, rather than being sold privately or sent off to Kanchi (Mahalingam 1989, 334, Cg.-1305). This suggests that had it not been specified otherwise, Tirumukkūṭal's local produce would likely have been traded in Kanchi, simply as a matter of default. Another part of the inscription designates allotments of water for elephants, horses, and other animals that may have come passing through (Mahalingam 1989, 334, Cg.-1305). We get the sense that Tirumukkūṭal was a small but well-frequented "river city," where the flowing water was integral to the culture, economy, and ethos of the place.

Similar inscriptions can be found on temples in other areas surrounding Kanchi, places we will now turn to in greater detail. The constellation of temple sites around the city in all directions follow the course of rivers that no longer flow (Figure 3.2). Each of these sites functioned as a settlement of village- or town size that sustained significant interactions with other sites in its vicinity, as well as with the urban hub. These places all have ancient histories as settled areas, but they flourished as multireligious economic centers during the eighth through thirteenth centuries. Like Tirumukkūṭal, most of these places constitute little more than remote villages that are for the most part outside of enduring pilgrimage circuits. The presence of at least one stone temple or rock-cut shrine that was constructed during the Pallava or the Chola period, and that subsequently received endowments and modifications, stands as testament to each area's former prosperity. Lush paddy fields surrounding the temples in each of the rural sites indicates functioning, highly engineered, ancient irrigation networks. In these places, we find shared patterns in architecture, iconography, and inscriptions, as well as shared uses of agricultural technologies, that weave together a network of riverside temple sites.

South of Kanchi: Arpakkam and Mākaṟal

In the fertile tract of land between the Palar and Ceyyar rivers, villages such as Arpakkam and Mākaṟal boast important monuments built in the Chola period. Mākaṟal is dominated by a large-scale temple dedicated to Shiva, with apsidal walls very finely carved in bold, high, relief (Figure 3.7). According to its architectural style and the dates of its earliest inscriptions, this temple was constructed in the middle of the twelfth century. Although now dominated by Hinduism, Mākaṟal was a multireligious and multisectarian place. Across from the Shiva temple is a Vishnu temple of indeterminate date. The remnants of a Jain establishment stand to its west, and a Goddess shrine is reportedly a kilometer outside the village.[15] Although inscriptions record that Mākaṟal was considered the locality's "main place" (*nāṭu*), its associated *tani-ūr* of Arpakkam also seems to have risen to a notable level of prominence during the eleventh through thirteenth centuries. Like Mākaṟal, Arpakkam contains a cluster of temples and quarries, surrounded by verdant paddy fields.

Perhaps because it is positioned not directly along the main road but a kilometer farther east, Arpakkam's former multireligious composition, as well as the richness of its natural resources, remain more apparent than they are at Mākaṟal. Within the space of well under a square kilometer, Arpakkam has four extant temples dedicated to Shiva, Vishnu, the Goddess, and the Jina Ātinātha.[16] Built of the local charnockite granite, all of these temples have architectural forms, sculptures, and inscriptions ranging from the eleventh through thirteenth centuries. Inscriptions dated in the regnal

15 Photographs in the archives of the Institut Français de Pondichéry and AIIS show a set of seven stone figures of goddesses (*sapta kaṉṉikaḷ*) filed under Mākaṟal. Residents report the shrine is about a kilometer outside the village, but I have not located it personally.

16 For discussion of Arpakkam's individual temples, and the renovations each has undergone, see Stein (2017, 123–131).

Figure 3.7 Mākaṟaleśvara (Shiva) temple, in the village of Mākaṟal, temple constructed ca. mid-twelfth century (Photo by Emma Natalya Stein)

years of the most famous Chola kings form veneers of words across the walls. On the southwest side of the Shiva temple, an extensive inscription from circa 1168 tells us that Arpakkam was even considered so important as to be selected as a gift of gratitude to a Chola general for his heroic acts in a major battle between the Cholas and the army of Ceylon (present-day Sri Lanka).[17]

Additional archaeological evidence, from as early as the megalithic period through the early second millennium CE, can also be found in the village and surrounding area.[18] Along the small road to the village is a ruined brick shrine embellished with stone and containing fragmentary granite sculptures of goddesses. A sculpture of the Buddha, datable stylistically to the eleventh/twelfth century, and a now half-buried stone slab depicting the goddess Jyeṣṭā are decontextualized remnants of Arpakkam's diverse cults of worship.

Arpakkam's prosperity very likely had to do with its ideal location between the two rivers. In addition to arable lands for rice cultivation, it also was set amid extensive stone quarries that made the construction of temples economically efficient (Figure 3.8). As the preferred material for elite

17 *ARE* (1899, no. 20); *SII* 6 (no. 456).
18 Local cairn circles are reported in *IAR* (1978–1979, 21).

Figure 3.8 Remains of a stone quarry, Arpakkam Village (Photo by Emma Natalya Stein)

sacred architecture, stone was a valuable commodity in South India. The availability of stone may even have led to the establishment of a workshop of stonemasons living in or near to Arpakkam village. Quality of craftsmanship is high. For example, at the apsidal-shaped Shiva temple—itself a virtuosic design—is a now loose sculpture of the Tamil deity Ayyaṉār so refined that it could only have been produced by expert artisans under substantial sponsorship. If not living in the village itself, the artists would have been specially brought in for building the temples and sculpting their walls and iconic deities.

Despite its once renowned status, Arpakkam has become all but forgotten, both in scholarly literature and in most devotional circuits. With the exception of the Jain temple, Arpakkam's sacred edifices remain locked except for single hours in the evening and very occasional mornings, when local villagers pay homage to the township's gods.

West of Kanchi: Kāvērippākkam

Villages that are similarly obscure can be found along the Palar River to Kanchi's west. For example, the village of Kāvērippākkam (25 kilometers west of Kanchi) gives us the most striking example of a place that once garnered both royal and military attention but subsequently has been all but erased.

Colonial-era records tell us that a large fortress was constructed there in the nineteenth century. Kāvērippākkam was the battlefield for several significant clashes between the Mysore sultanate, led by Tipu Sultan, and the forces of the British colonial empire. In his firsthand account, the Viscount Valentia noted damage that Tipu Sultan had inflicted on Kāvērippākkam's artificial lake, which was known to the author as the largest reservoir in southern India (Valentia 1811, 396). Remarks in this and similar colonial literature suggest that a journey towards Kanchi's west could present an especially significant challenge, in terms of climate and political volatility.

Kāvērippākkam's earlier historical importance can been gleaned through scattered material evidence. Its vast artificial lake has a bund of five kilometers on its eastern side, through which to water the surrounding agrarian lands. The long bund survives as evidence of the remarkable labor force that constructing so large a lake would have required. The Madras Government Museum houses a sizable collection of sculptures and architectural materials from the village. These fragments appear to have originally comprised at least two large-scale temples dedicated to Shiva. All are stylistically datable to the ninth century and are made of the dark, basalt-like stone called metagabbro.[19]

The fabric of the two Shiva temples can no longer be found on site. The structures must have been destroyed, or dismantled and removed, at an unknown point in time. However, a group of disengaged sculptures very similar to those in the Madras Government Museum stands on the premises of Kāvērippākkam's Kōṅkaṇēsvara temple, the structure of which post-dates the sculptures by several centuries. Although temples often get repaired or rebuilt, in Tamil Nadu they typically retain the same basic footprint. The more recent temple therefore may reveal the location of at least one of the original monuments. In addition to the sculptures *in situ* and in the museum, a large, now-dispersed group of monumental sculptures of yoginis also very likely was produced in Kāvērippākkam.[20] Now a minor locality, Kāvērippākkam must have been a site of some economic and cultic prominence around the year 1000.

19 Petrographic analysis by Conservation Scientist Janet Douglas on a sculpture in the Arthur M. Sackler Collection (S1987.905) revealed the type of stone. CSR Object Records, LRN 4268, 22 June 2012, National Museum of Asian Art, Smithsonian. An alternate term used by geologists for this rock is basic granulite.

20 Padma Kaimal has written extensively about the yogini group. She refers to them as the Kanchipuram yoginis. However, the similarity in style and type of stone of the fragments *in situ* and in the Madras Government Museum leaves little doubt that the yoginis are also from Kāvērippākkam. For discussion of provenance, see Kaimal (2012) and Kasdorf and Stein (forthcoming 2022).

North of Kanchi: Tiruttaṇi

Unlike the sites already discussed (Kāvērippākkam, Arpakkam, and Tirumukkūṭal), the village of Tiruttaṇi (40 kilometers to Kanchi's north) remains a bustling devotional thoroughfare. Today Tiruttaṇi houses four structural temples, including one for Shiva, one for Vishnu, and two for the popular Tamil deity Subrahmāṇyam, one of which stands atop a large hill. A shrine for the Goddess can also be found among the cluster of temples situated in the low-lying banks of a now-dry river.

The temple of greatest interest from an archaeological standpoint is the ninth-century Shiva shrine. This temple is important in that it anticipates the early Chola mode of construction that would dominate Tamil Nadu for at least the next two centuries.[21] It does so in three main ways:

(1) The Shiva temple is one of the rare shrines to bear a foundation inscription. The record tells us that it was built by a local dignitary named Nambi Appi.[22]
(2) This is the earliest surviving temple in the region to be constructed completely of granite (called *karuṅkallāl*, "black stone," in the inscription). Earlier temples were composed of sandstone, or more typically of a granite base with brick walls and brick superstructure.
(3) Unlike the eighth-century sandstone temples whose walls were replete with relief carvings, Tiruttaṇi's Shiva temple carries only one image on each exterior wall (Figure 3.9).

The identity of these deity images serves as an important connector among a remarkably widespread group of temples, in which the icons are positioned according to the cardinal directions towards which they face. The elephant-headed god Gaṇeśa and Shiva as the teacher Dakṣiṇāmūrti occupy the southern wall of the *ardhamaṇḍapa* (entry hall) and *vimāna* (main shrine) respectively, Vishnu is on the rear wall (west for an east-facing temple, east for a west-facing temple), and Brahmā and the Goddess are placed on the north *vimāna* and *ardhamaṇḍapa* respectively. Temples built during the ninth through thirteenth centuries typically bear this iconographic program. This holds true not only throughout Tamil Nadu, but also much

21 The features are pointed out in Gillet (2016, 451–452). On the early Chola style, see Kaimal (1996). For architectural description of the Virāṭṭaṉeśvara temple, see K.R. Srinivasan (1983a, 102–103).

22 Gillet convincingly revises earlier interpretations of the inscription, which had attributed the temple to a king (Gillet 2016, 448).

Figure 3.9 Virāṭṭaṉeśvara (Shiva) temple, Tiruttaṇi Village, temple constructed ca. ninth century (Photo by Emma Natalya Stein)

farther afield. In Polonnaruwa, Sri Lanka, migrant merchant communities who built Chola-style temples selected this particular configuration of deities for the walls.

However, Tiruttaṇi is best known not for the Shiva temple, but for the hilltop shrine dedicated to Subrahmāṇyam. As Gillet has discussed, although a shrine existed on the hill by the ninth century, nothing of its original construction survives in the present temple-complex (Gillet 2016, 457–458). The centuries of reconstruction and the layers of donation inscriptions reveal that the site has been active continuously. The fact that the shrine for Subrahmāṇyam—rather than Shiva—gained such traction has to do with the shifting religious landscape in post-Chola Tamil Nadu. However, the factors that contributed to the survival of any site in the now remote village of Tiruttaṇi are also worth exploring. The site may have maintained its status in part because of its unique geographic position. Tiruttaṇi is situated approximately midway between two major ritual centers, precisely along the road that connects them. Forty kilometers to the south is the city of Kanchi; 60 kilometers north is the "golden temple" at Tirupati. Tirupati's hilltop shrine remains the most holy destination for devotees of Vishnu in South India, and it is one of the most prosperous religious establishments in the entire subcontinent.

The relationship with Tirupati and the pilgrimage route seems to have been an important factor in Tiruttaṇi's early development. Gillet has pointed out an early eleventh-century inscription on the Shiva temple that ensures the provision of food for pilgrims coming from or going to Tirupati (Gillet 2016, 476). As she concludes, the Shiva temple must have stood directly along the road, and the road must have already been a thoroughfare by the date of the inscription. To further this point, she also persuasively suggests that the Vishnu temple currently standing in Tiruttaṇi was relocalized from a nearby village in order to place it nearer to the road (Gillet 2016, 472).

Tiruttaṇi's position as a convenient stop between Kanchi and Tirupati probably contributed to the site's level of continuity. Connections between Kanchi and Tirupati remain strong in present-day devotional circuits. Groups of pilgrims visit Kanchi's temples on the way to or from Tirupati. Some wear yellow garments to announce that the golden temple is their destination. Others have shaved heads, which shows they are returning from Tirupati, where hair is an esteemed sacred offering. Tiruttaṇi's position along this road, and its affiliation with its own pilgrimage place (the Subrahmāṇyam temple on the hill), seems to have enabled it to maintain a level of prominence not shared by other villages in the area.

Tiruttaṇi's proximity to a river may have further enabled it to endure. The village spans the banks of the Nandi, a branch of the Kusasthalai river. Beginning far to the north, the Kusasthalai connects the state of Andhra Pradesh with Tamil Nadu. It once eventually watered the large lake at Kāvērippākkam, and then flowed into the Palar River just east of Arcot.[23] A series of villages containing Pallava- or Chola-era temples also lie near the Kusasthalai and Nandi rivers. However, Tiruttaṇi is unique in its position at the juncture of the Nandi River and the Tirupati road.

Unfortunately, like so many of Tamil Nadu's rivers, today the Kusasthalai and the Nandi rivers are dry. A large dam on the southern edge of the Ammapalli reservoir in Andra Pradesh constricts the Kusasthalai's waters before they can even enter Tamil Nadu.

Disappearance and Decline

The Palar River still flowed—at least in some years—during the last quarter of the nineteenth century, when Kanchi's district collector, Charles Stewart Crole, reported that it "continues in fresh, more or less high, for six months on end" (Crole 1879, 2). Although the river had been dry from 1867 to 1870, in 1874 Crole wrote that when it flows,

> Owing to the rapid fall in its course, the current of the Palar is very strong. This, and the fact of the sand in its bed for some feet down being in a state of motion, render crossing this river, even during ordinary floods, an operation of considerable difficulty, resulting occasionally in loss of life. (Crole 1879, 2)

Although Crole's report tells us that the Palar was a mighty river a mere century ago, today it is dry. Within Kanchi, the Palar's tributary, the Vegavathi, is little more than a polluted drainage canal (Figure 3.10). A walk through the southern belt of the city brings us to this stagnant remnant, its surface masked by green moss pocked with intermittent plastic bags, bottles, and other forms of refuse. It is hard to imagine that this could be all that remains of the Vegavathi, that river the poets once described as a source of joy (Cēkkiḻār 2006, 109–110).

23 Another offshoot of this river is visible in Figure 9.1 of this volume, which shows rivers in and around the city of Madras/Chennai.

Figure 3.10 Vegavathi river, Kanchi, today a drainage canal for the city (Photo by Emma Natalya Stein)

The health of the river is a measure of the health of the city on its banks. When water flowed through northern Tamil Nadu, Kanchi and the suburban settlements that surrounded it were elite places trafficked by diverse communities who used and coexisted with the rivers. This constellation of sacred and economic centers was important on the regional level. But it was also important transregionally. Inscriptions and iconographic patterns show us that the river's pathway to the sea not only facilitated contact between the urban core and the suburban hinterlands. The roads that led across this expanse of land also gave the city access to the web of extended networks upon which the premodern global economy relied.

Kanchi itself was known to be a gateway to the seas. The Chinese biography of the Buddhist pilgrim Xuanzang tells us Kanchi was a place of refuge for Sri Lankan monks, and that it was the port of embarkation for travel to Sri Lanka (Hwui Li 1911, 138–139). Just outside Kanchi, an evocative inscription preserves more evidence of connections with Sri Lanka. Elegantly ornamenting a full wall of Arpakkam's Shiva temple, this twelfth-century record tells us that a military general celebrated his victory in a naval battle with Sri Lanka by gifting this fertile village to the temple's god in honor of his guru.[24]

24 *ARE* (1899, no. 20); *SII* 6 (no. 456).

Meanwhile, migrant Tamil merchant communities were settling in cities overseas. In distant places, such as Polonnaruwa, Sri Lanka, they sponsored the production of local temples. Although built by Sri Lankan stonemasons, these temples reveal the patrons' preference for their own familiar forms—a square sanctum with a modest pyramidal superstructure and flat-roofed entry hall, each wall inset with a single niche containing an icon of a deity, and no further ornament besides pilasters carved in shallow relief. This architectural formula was exported from the Tamil homeland.

Iconography was relatively consistent as well. The arrangement of deities corresponded to the direction they faced. Beginning in the tenth century, a standardized program of deities on temple walls was transmitted throughout Tamil Nadu and across its diaspora. Other Tamil elements were incorporated in places as far afield as Cambodia. In a relief carving at the tenth-century temple of Banteay Srei, the female ascetic Kāraikkaḷ Ammāyar is shown playing cymbals to accompany Shiva's dance. Near unique within Cambodia, this tableau was common in Tamil Nadu. The widespread consistency of iconography is one of many examples of broader circulations between South and Southeast Asia.

For the full millennium spanning Kanchi's Pallava foundations in the third century to the fall of the Cholas in the thirteenth, the city endured a relatively stable status as a royal capital and a prominent urban center. As such, it served as a natural hub in the transmission of objects, ritual practices, and systems of knowledge. The lives of rivers and cities in Kanchi's hinterlands are historically intertwined and mutually dependent. Where rivers flowed, settlements emerged as important nexuses of regional and transregional travel and trade. Kanchi's rivers have unfortunately suffered the same fate as rivers throughout the Indian subcontinent. Largely as a result of overpopulation and deforestation, as well as dam construction, erosion, drought, global climate change, and a greater focus on industrial rather than agrarian productivity, their waters have incrementally dried up.[25] Consequently, the agrarian-based centers boasting important temples on their banks have gradually deteriorated and all but disappeared. Now situated beside dry, sandy beds, the once-prosperous places in Kanchi's hinterlands have become difficult even to find.

25 My thanks go to Professor Narpat Singh Rathore (Geology department, M.L. Sukhadia University) for sharing his insights into the drying of India's rivers at the workshop in Udaipur, Rajasthan, 6–7 March 2019 (see footnote 1).

References

Annual Reports on Indian Epigraphy [*ARE*]. 1887–1905. Janpath: Archaeological Survey of India.

Cēkkiḻār. 1950. *Periya Puranam*. C.K. Subiramaniya Mukaliyar.

Cēkkiḻār. 1990. *St. Sekkizhar's Periya Puranam* (T. N. Ramachandran, Trans.; Vol. 1). Tanjavur: Tamil University.

Cēkkiḻār. 1995. *St. Sekkizhar's Periya Puranam* (T. N. Ramachandran, Trans.; Vol. 2). Tanjavur: Tamil University.

Cēkkiḻār. 2006. *The History of the Holy Servants of the Lord Shiva: A Translation of the Periya Purāṇam of Cēkkiḻār* (A. McGlashan, Trans.). Crewe: Trafford Publishing.

Champakalakshmi, R. 1993. "The Urban Configurations of Toṇḍaimaṇḍalam: The Kāñcīpuram Region, c. A.D. 600–1300." In *Urban Form and Meaning in South Asia: The Shaping of Cities from Prehistoric to Precolonial Times*, edited by D.M. Srinivasan and H. Urban, 185–207. Washington, DC: National Gallery of Art.

Crole, Charles Stewart. 1879. *The Chingleput, Late Madras, District: A Manual Compiled Under the Orders of the Madras Government, by Charles Stewart Crole, of the Madras Civil Service*. Kolkata: Lawrence Asylum Press.

Dehejia, Vidya. 1988. "The Persistence of Buddhism in Tamil Nadu." In *A Pot-Pourri of Indian Art*, edited by P. Pal, 53–74. Mumbai: Marg Publications.

Dehejia, Vidya. 2016. *The Thief Who Stole My Heart: The Material Life of Chola Bronzes from South India, c. 855–1280* A.W. Mellon Lectures in the Fine Arts, National Gallery of Art, Washington, DC. Available at: http://www.nga.gov/content/ngaweb/audio-video/video/mellon2-2016.html, accessed 20 January 2022.

Digital Tēvāram/Kaṇiṉit Tēvāram [CD-Rom]. 2007. Edited by Jean-Luc Chevillard and S.A.S. Sharma. Pondicherry: École Française d'Extrême-Orient/Institut Français de Pondichéry.

Foote, Robert Bruce. 1865. "On the Occurrence of Stone Implements in Lateritic Formations in Various Parts of the Madras and North Arcot Districts." *Geological Magazine*, 2(17): 503–504.

Foote, Robert Bruce. 1916. *Prehistoric and Protohistoric Antiquities of India*. Delhi: Leeladevi Publications.

Francis, Emmanuel. 2014. "Royal and Local Bhakti under the Pallavas." In *Mapping the Chronology of Bhakti: Milestones, Stepping Stones, and Stumbling Stones*, edited by V. Gillet, 97–134. Pondicherry: Institut Français de Pondichéry and École Française d'Extrême-Orient.

Gillet, Valérie. 2010. *La création d'une iconographie Shivaïte narrative: Incarnations du dieu dans les temples pallava construits*. Pondicherry: Institut Français de Pondichéry, and Ecole Française d'Extrême-Orient.

Gillet, Valérie. 2016. "Gods and Devotees in Medieval Tiruttaṇi." In *The Archaeology of Bhakti II: Royal Bhakti, Local Bhakti* (Collection Indologie no. 125), edited by E. Francis and C. Schmid, 443–494. Pondicherry: École Française d'Extrême-Orient/ Institut Français de Pondichéry.

Heitzman, James. 1987. "Temple Urbanism in Medieval South India." *The Journal of Asian Studies*, 46(4): 791–826.

Heitzman, James. 1997. *Gifts of Power: Lordship in an Early Indian State*. Oxford: Oxford University Press.

Heitzman, James. 2001. "Urbanization and Political Economy in Early South India: Kāñcīpuram During the Cōḻa Period." In *Structure and Society in Early South India: Essays in Honour of Noboru Karashima*, edited by K.R. Hall, 117–156. Oxford: Oxford University Press.

Heitzman, James, and S. Rajagopal. 2004. "Urban Geography and Land Measurement in the Twelfth Century: The Case of Kanchipuram." *Indian Economic and Social History Review*, 41(3): 237–268.

Hultzsch, E. (ed.). 1992. *South Indian Inscriptions, Vol. 2, Part 3: Tamil Inscriptions in the Rajarajesvara Temple at Tanjavur and Other Miscellaneous Records of the Chola, Pallava, Pandya and Vijayanagara Dynasties*. Janpath: Archaeological Survey of India.

Hwui Li. 1911. *The Life of Hiuen-Tsiang, by the Shaman Hwui Li*. (S. Beal, Trans.). London: Kegan Paul, Trench, Trubner.

Indian Archaeology: A Review [*IAR*]. 1953–2005. New Delhi: Director General, Archaeological Survey of India.

Kaimal, Padma. 1996. "Early Cōla Kings and 'Early Cōla Temples': Art and the Evolution of Kingship." *Artibus Asiae*, 56(1/2): 33–66.

Kaimal, Padma. 2012. *Scattered Goddesses: Travels with the Yoginis*. Ann Arbor, MI: Association for Asian Studies.

Kasdorf, Katherine and Emma Natalya Stein. Forthcoming 2022. "Alternate Narratives for the Tamil Yoginis: Reconsidering the 'Kanchi' Yoginis Past, Present, and Future." In *Religions. Scattering and Destroying: On the Unforeseen Consequences of Collecting and Reuse in South Asian Art*, edited by Janice Leoshko, Padma Kaimal, and Catherine B. Asher.

Mahalingam, T.V. 1969. *Kāñcīpuram in Early South Indian History*. Hyderabad: Asia Publishing House.

Mahalingam, T.V. (ed.). 1989. *A Topographical List of the Inscriptions in the Tamil Nadu and Kerala States: Chingleput District*, Vol. 3. New Delhi: Indian Council of Historical Research.

Michell, George (ed.). 1993. *Temple Towns of Tamil Nadu*. Mumbai: Marg Publications.

Minakshi, Cadambi. 1938. *Administration and Social Life Under the Pallavas*. Kolkata: University of Madras.

Orr, Leslie. 2004. “Processions in the Medieval South Indian Temple: Sociology, Sovereignty and Soteriology.” In *South-Indian Horizons: Felicitation Volume for Francois Gros on the Occasion of his 70th Birthday*, edited by J.-L. Chevillard and E. Wilden, 437–470. Pondicherry: Institut Franç ais de Pondichéry and École Franç aise D’Extreme-Orient.

Orr, Leslie. 2006. “Preface.” In *Pondicherry Inscriptions*, edited by B.S. Kuppusamy and G. Vijayavenugopal, Vol. 1, i–xxvii. Pondicherry: Institut Français de Pondichéry and École Française d’Extrême-Orient.

Pattuppāṭṭu. 2012. [The Ten Tamil Idylls] (A. Dakshinamurthy, Trans.). Kattankulathur: Tamil Perayam, SRM University.

Prapañca, Mpu. 1995. *Deśawarṇana (Nāgarakṛtāgama) by Mpu Prapañca* (Stuart Robson, Trans.). (Verhandelingen van het Koninklijk Instituut voor Taal–, Land– en Volkenkunde 169). Leiden: KITLV Press.

Shattan. 1993. *Manimekhalai* [The Dancer with the Magic Bowl] (A. Danielou and T.V. Gopala Iyer, Trans.). Delhi: Penguin Books.

Srinivasan, C.R. 1979. *Kanchipuram through the Ages*. Delhi: Agam Kala Prakashan.

Srinivasan, K.R. 1983a. “Pallavas of Kāñcī: Phase I.” In *Encyclopaedia of Indian Temple Architecture*, Vol. 1, Part 1, edited by M.W. Meister and M.A. Dhaky, 87–106. New Delhi: American Institute of Indian Studies.

Srinivasan, K.R. 1983b. “Pallavas of Kāñcī: Phase II.” In *Encyclopaedia of Indian Temple Architecture*, Vol. 1, Part 1, edited by M.W. Meister and M.A. Dhaky, 23–80. New Delhi: American Institute of Indian Studies.

Stein, Burton. 1980. *Peasant State and Society in Medieval South India*. Oxford: Oxford University Press.

Stein, Emma Natalya. 2017. *All Streets Lead to Temples: Mapping Monumental Histories in Kanchipuram, ca. 8th–12th centuries CE*. PhD Dissertation. Yale University.

Stein, Emma Natalya. 2021. *Constructing Kanchi: City of Infinite Temples*. International Institute for Asian Studies, Asian Cities Series. Amsterdam: Amsterdam University Press.

Subrahmanya Aiyer, K.V. (ed.). 1928. *South Indian Inscriptions* [SII], *Vol. 6: Miscellaneous Inscriptions from the Tamil, Telugu, and Kannada Countries*. Archaeological Survey of India,

New Imperial Series. Madras: Archaeological Survey of India.

Valentia, George Annesley Viscount. 1811. *Voyages and Travels to India, Ceylon, the Red Sea, Abyssinia, and Egypt in the Years of 1802, 1803, 1804, 1805, and 1806*, Vols. 1 (Text) and 4 (Plates). London: Rivington.

Venkataraman, K.R. 1973. *Dēvī Kāmākshī in Kāñchī: A Short Historical Study*. Colombo: Sri Vani Vilas Press.

Xuanzang. 1884 [reprint by Munshiram Manoharlal 1969]). *Si-yu-ki. Buddhist Records of the Western World. Translated from the Chinese of Hiuen Tsiang (A.D. 629) by Samuel Beal*, Vol. 2. London: Trubner & Co.

Biography

Emma Natalya Stein received her PhD in Art History from Yale University (2017), and is Assistant Curator of South and Southeast Asian Art at the National Museum of Asian Art, Smithsonian. She regularly conducts fieldwork on the intersection of art and landscape, and lectures in Asia, Europe, and the USA.

4 The Political Economy of Banjarmasin's River Landscape during the Sultanate Period (1526–1860)

Vera D. Damayanti

Abstract

Banjarmasin, the capital of South Kalimantan Province, is known in Indonesia as the city of a thousand rivers. Situated on a fluvial plain and close to the mouth of the Barito River, Banjarmasin's landscape was transformed as a result of interaction between the physical environment and human actions. This chapter explores the landscape changes in Banjarmasin during the Sultanate period (1526–1860) that came about because of warfare, contacts with foreign traders, and the division of the territory under the treaties with the Dutch. The creation of space for political, economic, and settlement functions combined with the riverine and tidal swamp environment defined the development of the city, making it different from the other coastal port cities of Indonesia.

Keywords: landscape transformation, riverscape, political economic influence, Banjarmasin Sultanate

Introduction

Banjarmasin, the current capital of the province of South Kalimantan in Indonesia, is situated close to the mouth of the Barito River. It is known as "the thousand river city" because of the immense number of rivers, streams, creeks, and canals that flow through the city. Moreover, interaction between the river and the city's inhabitants is still visible in contemporary urban life, creating a sharp image of Banjarmasin as a river city (see Figure 4.1).

Padawangi, Rita, Paul Rabé and Adrian Perkasa, *River Cities in Asia.*
Taylor & Francis Group, 2022
DOI: 10.5117/9789463721851_CH04

Figure 4.1 The heritage of cultural landscape which among others can still be found in Banjarmasin city today (a) pole houses along the Kuin River and (b) traditional floating market at the mouth of the Kuin River, Banjarmasin (Photos by Vera D. Damayanti)

Looking back to the "age of commerce" between the fifteenth and seventeenth centuries in Southeast Asia, Banjarmasin was one of a number of port cities that played a role in the maritime trade network (Reid 1988).[1] It appears that its advantageous geographical position within the commercial network, as well as its being recognized as a pepper producer—at that time a commodity in high demand in the Western market—turned this city-state into a prominent entrepôt in the seventeenth century. At the same time, Banjarmasin was put in a precarious position, a position which will be discussed in this chapter. In fact, the political economy of the pepper trade contributed to a predicament that lasted more than three centuries. It also brought about the human-landscape interaction that engendered the *longue durée* of landscape transformation, which in turn affected the development of the urban landscape of this riverine port city.

This chapter discusses the transformation of the landscape of Banjarmasin during the Sultanate period (1526–1860) and explores the relationship between cities and rivers from a historical perspective. This period is important since it became the fundamental layer of Banjarmasin's landscape history in political, social, cultural, and economic contexts in the early stages of the formation of the city. The end of the Sultanate period in 1860 was followed by the transfer of power to the Dutch; subsequently, Banjarmasin entered the colonial period, bringing the city into another stage of landscape transformation. This involved dynamic processes that further shaped Banjarmasin into a colonial town with a distinctive character. This chapter is limited to the change in landscape during the Sultanate period when Banjarmasin was still perceived as a traditional city.

In order to reconstruct the past of Banjarmasin's landscape, the research is based on historical data collected from archival and published sources consisting of written, illustrated, and cartographic materials. One of the most important records pertaining to the early period of the sultanate in the fifteenth and sixteenth centuries is the Banjarmasin Court Chronicle which was extensively studied by Johannes Jacobus Ras (1968) in his dissertation *Hikajat Bandjar*. In order to reconstruct the history of the landscape in this chapter, it is analyzed through the lens of the political economy of the pepper

1 The "age of commerce" is defined in Southeast Asia as the period between the fifteenth and seventeenth centuries when trade grew enormously in the region. This growth was stimulated by the economic transformation that prevailed in mainland Asia—including China, India, Persia, and the Arab Peninsula—that encouraged traders from these places to venture into the region. This flourishing of trade considerably influenced the construction of the complex maritime trade network in Southeast Asia that involved local port cities that were mostly located on the coast and estuaries and connected to rivers (see Reid 1988).

trade, which had a remarkable influence on the life of the Sultanate of Banjar. The changes in space and relevant landscape elements are investigated to formulate the phases of landscape evolution. The influence of the river on the city is identified mainly through the roles the river played in its associations with political and economic interests. Indeed, the historical processes that affected the development of the Banjarmasin landscape provide a better understanding of the relationships between rivers and city life and are also useful in terms of comprehending the origin of the urban pattern that exists in the city today.

The Emergence of Southeast Asia's Port Cities

Before turning our attention to Banjarmasin, we first consider the origin of the city in Southeast Asia as a point of departure for uncovering the landscape of this city in the early stages of its establishment. Terence G. McGee (1969) asserted that the growth of indigenous cities in the region was influenced by the trading contacts in the first century that introduced Chinese and Indian religions and systems of political organization.

The political, economic, and technological developments that took place in Southeast Asia during the fifth century caused the character of settlement centers in this region to diverge. In terms of economic and cultural factors these could be categorized as the sacred and the market cities. This divergence nonetheless was influenced by, among others, the geographic features and ecological conditions that impacted on the economic base of each area (Sutherland 2007). The sacred cities were located inland with an agriculture-based economy—for example, Angkor Thom, which was established in the early ninth century by the Khmer Empire in Cambodia. The market cities, on the other hand, were usually situated on the shorelines or river mouths and relied on trade in order to gain wealth. As a consequence of these economic activities, the market cities became port cities—for instance, Srivijaya on the east coast of the island of Sumatra was a port city that was considered the most significant of the early port cities, reaching the height of its power in the eighteenth century (McGee 1969). Trade among the states of mainland Asia such as China, India, Persia, and with Arabs and Malays flourished in the fifteenth century, resulting in an increase in trading volume as well as the creation and expansion of the maritime trade network. This phenomenon played a significant role in the growth of port cities as the hubs of the traders' journeys and, subsequently, strengthening of the network in the region (Reid 1980; Chafee 2008).

During the age of commerce, Southeast Asia's port cities demonstrated a character different from the neighboring cities on mainland Asia and their Western counterparts. Previous studies have confirmed that those market cities had a similar pattern with regard to economic, social, cultural, and political features and the physical environment.[2] The dual function—as the political center and primary marketplace—is the primary attribute of the port city. This entails the port-polity (Kathirithamby-Wells 1990; Tagliacozzo 2007). Herein, two landscape elements become important landmarks: the royal palace and the port-of-trade of the local state, which were generally located in close proximity to each other (Kathirithamby-Wells 1990). The royal palace served as the primary landmark and signified the center of power. While the ruler, his/her family, and the royal elites commonly lived in and adjacent to the royal palace, the commoners and traders who were clustered in groups based on their origin and ethnicity dwelled in areas outside the royal palace and distant from the royal elite (McGee 1969). If a ruler controlled the royal palace, a *syahbandar*[3] ("harbourmaster") managed and promoted the port on behalf of the ruler.

With market and port cities located on river mouths and in coastal areas, this condition had an impact on the ways in which the distinct characteristics of the Southeast Asian cities were shaped. Commonly situated in favorable places providing shelter for foreign vessels and commercial exchange, these port cities became gateways through which commercial commodities could be channeled and were a suitable meeting point for traders from inland and overseas. Those traders who lived in ethnicity-based compounds created a cosmopolitan and pluralist city in the social, cultural, and political context. However, if port cities were located in a fluvial environment, the cities encountered physical environmental problems due to unfavorable soil and the hydrological conditions of a swampy terrain that had profound influences on the population and town development. Firstly, the inhabitants' income came from trade and their subsistence from imported foods and other staples; secondly, they lived in houses built on wooden stilts or floating houses. Furthermore, the swamp environment prevented these port cities from constructing walled cities and houses on land similar to those usually found in the cities of mainland Asia. As a consequence of there being an

2 See Bronson (1977); Tagliacozzo (2007, 911–932); Reid (1980, 235–250).

3 Since these commercial centers were open to international commerce, a *syahbandar* was usually a foreigner who had good knowledge of the languages and cultures of the foreign traders; see: Tagliacozzo (2007, 914); Reid (1993, 120).

abundance of green space, Southeast Asia's port cities had the appearance of a suburban landscape, which in fact was suitable for overcoming the heat of the tropical sun and to take advantage of the breeze to reduce humidity (Reid 1980).

This outline of the cities' features highlights the fact that the character of the landscape of Southeast Asia's port cities was a result of the combination of the physical environment and the activities of the population in the economic, social, cultural, and political contexts. This description also provides an insight into the significant correlation between humans and landscape since the physical landscape stimulated the population's economic interests which generated the port city, while the population, as the agent of landscape changes, shaped the performance of the city.

The Role of Banjarmasin in the Pepper Trade

The Sultanate of Banjarmasin was an Islamic kingdom, presumed to have been established in 1526 in South Kalimantan. This city-state was the successor of Nagara Dipa and Nagara Daha, two Hindu-Buddhist port-polities located further upstream, from which Banjar inherited Javanese traditional cultural practices (see Figure 4.2). Meanwhile, the Islamic influences came from the state of Demak, a paramount power in Java, which sent military assistance to Prince Samudera—also known as Sultan Suriansyah, the first ruler of the kingdom—when he was struggling against an attack by Nagara Daha shortly after the establishment of the sultanate. The court's hinterland territory was quite extensive and covered the southeast of Kalimantan and several parts of central Kalimantan.[4]

The political center of this city-state was initially located in Kampung Banjar, at the mouth of the Kuin River, a tributary of the Barito River, situated about 50 kilometers from the mouth of the Barito in the Java Sea (see Figure 4.2).[5] The landscape of this compound and surrounding area was predominantly tidal swamp, the result of the long-term geological process of the depositing of alluvial sediment in the Barito Basin during the Holocene period. The vast land of the back swamp area provided an unmissable opportunity for agriculture; however, it was unsuitable for farming unless a proper drainage system could be constructed. However, the swamp in Banjar and its vicinity remained

4 See *Hikajat Bandjar* by J.J. Ras (1968) for the establishment of the Sultanate of Banjarmasin.
5 Today this area is known as North Kuin.

Figure 4.2 Map indicating important locations related to the Banjarmasin Sultanate (Adapted by Vera D. Damayanti, 2019)

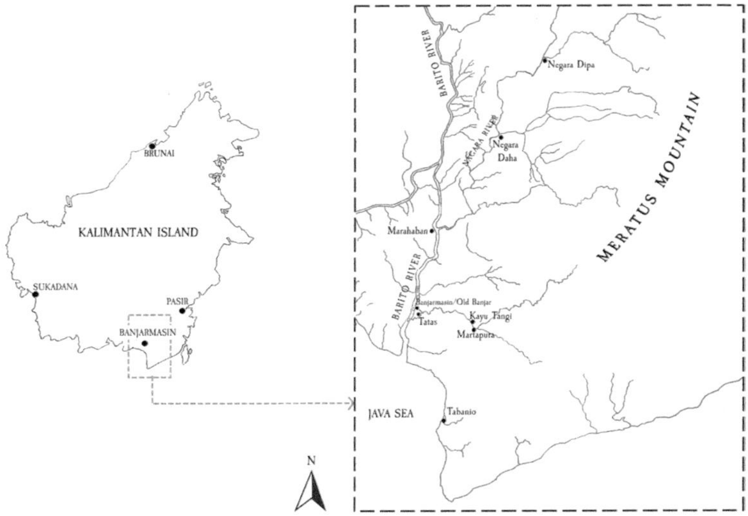

uncultivated during the Sultanate period, presumably due to the locals' lack of agricultural knowledge and labor (Knapen 2001). To meet food demand, the city imported rice from upstream and from Java. The city's economy was thus based on trading that involved much of the population, either as wholesalers, small retailers, or employees of the city's wealthy merchants. All commerce was controlled by the sultan through the institution known as the *Kyai Palabuhan*, the local title for the Malay *syahbandar* (Ras 1968).

Previously, Banjarmasin had been only an intermediary emporium in Java's vigorous ports-of-trade sea network. When the Javanese ports collapsed due to the expanding influence of the Mataram Kingdom in the middle of the seventeenth century, traders sought opportunities offered by a free trading port. Located just across from the island of Java, Banjarmasin accommodated those merchants who had decided to leave Java. This exodus of traders caused a shift of commercial activities to Banjar, which had a clear influence on the involvement of this entrepôt in the Southeast Asian commercial network. Another factor which drove the role of Banjar as a port-of-trade was that it became an alternate pepper producer since Jambi in Sumatera and Banten in West Java, the primary producers in the archipelago, could not provide enough pepper due to the monopoly of the pepper trade in Jambi and the blockade of Banten by the VOC (Suntharalingam 1963;

Ahyat 2012).[6] Banjarmasin also offered other exotic commodities extracted from the forests upstream, such as rattan, *damar*, wax, *gutta-percha*, edible birds' nests, fish, bezoar, camphor, dragon's blood (resin extracted from *Daemonorops draco* or rattan jernang), *Jambe* canes, diamonds, and gold (Beeckman 1718; Bock 1882; Knapen 2001).[7] These goods eventually resulted in the town becoming one of four main ports-of-trade on the island of Kalimantan besides Brunei in the north, Pasir in the east, and Sukadana on the west coast (Beeckman 1718).

Being known as a pepper producing region had a remarkable impact both on the economic context and on the sultanate's political affairs. Pepper attracted foreign merchants to trade in Banjarmasin, where the competing parties were predominantly Chinese, Dutch, and English merchants. Meanwhile, as the sultan and his royal family controlled the pepper commerce, internal disputes among these elites, triggered mainly by their economic interests, became inevitable. The sultan's dominance over the trade was contested by the heads of the districts, the hinterland rulers, as well as the Dayaks, the natives of Kalimantan. During this crisis, the Dutch and the English supported the sultan, which was agreed via treaties that were exchanged for the exclusive pepper trade.[8] However, these agreements gradually reduced the sultan's territory and undermined his sovereignty and later led to the dissolution of the sultanate in 1860.

Landscape Transformation during the Sultanate Period

During the sultanate period, the actors who engaged in this political-economic system were the sultan and his royal elite, local traders, foreign traders, and the people of Banjar. Power was initially held by the sultan

6 From this period on, the Sultanate became gradually increasingly recognized and mentioned in travel reports, mostly in connection with ship navigation and trade conditions. Another reason that Banjarmasin was favored in the pepper trade was because it was cheaper to buy pepper here than from other pepper producers such as Palembang on the island of Sumatera (see Beeckman 1718, 49).

7 Knapen, in his dissertation, studied extensively the source of commercial products of southeast Borneo within the period 1600–1880, and ascertained that the area was part of the Banjar sultanate's territory in the hinterland. The location of the sources—for instance Hulu Sungai for timber and non-timber products and Martapura for diamonds—indicates that the export commodities came from the upriver area.

8 See ANRI (1965) for numerous treaties between the sultan and the Dutch which were made between the years 1631–1856.

with regard to controlling the trade system, but as a consequence of shifting interests the sultan not only gained profit but, most importantly, obtained foreign military assistance to maintain the status quo. Power, however, was slowly transferred to these external entities. This process affected the Banjar riverscape, as is chronologically explained in this section.

The Port-Polity of Banjarmasin (1526–1612)

In 1526, Kampung Banjar, located on the mouth of the Kuin River, was designated a port-polity where a mix of political and economic functions would exist. These dual functions contributed to the urban pattern of this new emporium through the development of two main physical elements: the royal capital and the main trading port on the north bank of the Kuin surrounded by the swamp forest environs (see Figure 4.3).

The *keraton* ("royal court"), which symbolized the sultan's power, became the nucleus of the area. The court was configured by a building complex in a compound adjacent to the residence of the royal elites, religious elites, merchants, artisans and lay people, which collectively resulted in the concentration of a large population that created an urban center.[9] Since the local inhabitants perceived the ruler of the Southeast Asian port-polity city as a supreme master and spiritual leader, the complementary but imperative religious structure of a mosque was erected close to the court.[10] Presumably due to the limited width of the river levee, the riverbank was occupied exclusively by the royal elites—with the *keraton* and mosque—and state officers, while the rest of the population dwelled away from the riverbank in floating houses on the river (Beeckman 1718; Van Wijk 1951; Sumawinata 1998).[11]

Another key urban feature was the port-of-trade located close to the mouth of the Pelambuan River, a tributary of the Kuin River, which became the second center of the settlement (Ras 1968). It is presumed that the traders, which included the elite merchants of the *orang kaya* ("the wealthy

9 The *Hikajat Bandjar* describes the royal compound as consisting of *dalam agung* or *astana* ("the royal residence"), *paseban* ("courtyard"), and *sitilohor* ("the audience-hall"). These elements are similar to those found in Javanese royal palaces and it is likely that the Banjar court imitated the layout as they followed the Javanese tradition. The *Hikajat Bandjar* uses the local terms *pagar*, *pager*, or *sirap* to indicate the compound of the royal families (see Ras 1968, 362, 364, 574, 598).

10 The first mosque of the sultanate was presumed to have been on the current site of Masjid Suriansyah (see Irhamna and Safari 2001).

11 In his account, Beeckman depicted the condition of the new upstream capital, which he mentioned was similar to Tatas, the new downstream center in Banjar.

Figure 4.3 Reconstructed map of the Banjarmasin port-polity in its early establishment in 1526 (Adapted from Vera D. Damayanti, 2019)

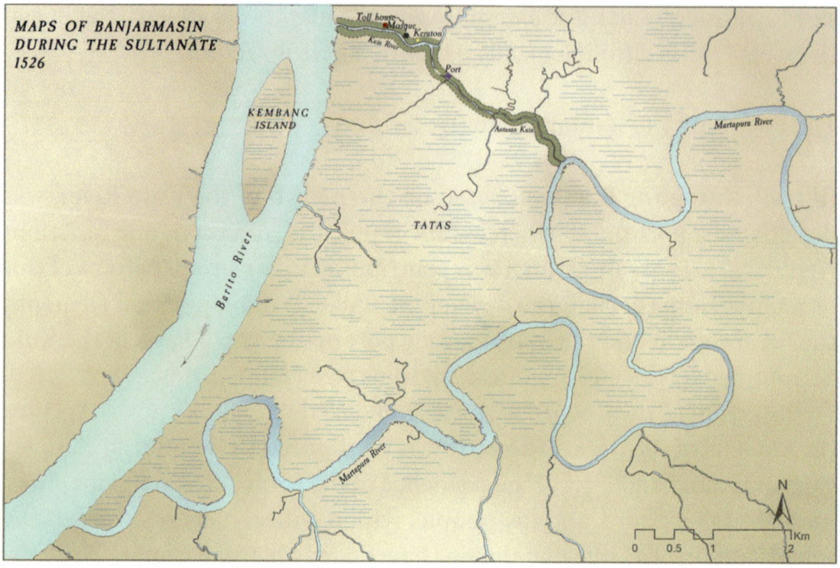

and powerful man"), resided close to the port, a situation similar to that observed in other Southeast Asian port cities (Reid 1980). Meanwhile, the existence of traders both insular—such as the Javanese and Bugis—and inter-regional—such as the Chinese and Arabs—accentuated the pluralistic character of the commercial center in Banjar. They lived together with the locals on the Kuin by creating enclaves based on their origin and ethnicity, where the population lived either in pole houses or on floating rafts (Beeckman 1718). A toll-house, which marked the city as a commercial center, was located in this area, yet it is hard to find sources indicating its exact location. In addition to the export trade that took place in the port, local trade was carried out by domestic retailers who sold various grocery products on small boats in the daily floating market on the Kuin River. These boats were usually managed by women while the men were involved in wholesale trading (Bock 1882; Reid 1993).

Like other Southeast Asian port cities, a semi-rural look characterized the visual landscape of Banjarmasin (Reid 1993, 88–89). The back swamp and riparian vegetation created the greenness described in Daniel Beeckman's account of his boat entering the mouth of the Kuin: "We had eight miles to go up this River, which is very crooked, where the scorching heat of the sun would have been as troublesome to us ...

and the trees wonderfully high on each side, we were pretty well shaded from the heat. We could not see the town 'till we were just ent[e]ring into it, because of the tallness of the trees that stand close together" (Beeckman 1718, 60).[12]

As the greenery provided a favorable microclimate for the inhabitants, the city-dwellers planted fruit trees such as coconut and banana on the river levee to improve their household economy and in order for them not to be completely dependent on the market. As outlined above it has been confirmed that in the initial stages of the development of the Banjar townscape, the influence of the political economy, which manifested itself through the establishment of the port-polity, shaped Kampung Banjar into a riverine-urban center typical of Southeast Asia's port cities.

Shifting the Capital: Banjarmasin as a Secondary Port City (1612–1663)

In the early sixteenth century, European traders became involved in the Southeast Asian maritime trade pioneered by the Portuguese in 1511. From then on, Europeans sent expeditions to the region in attempts to obtain exotic items, mainly spices. Among the Europeans, the Dutch and the English, two parties known as rivals in the trade market, frequently made contact with the Sultan of Banjarmasin. The Dutch established their first factory in Banjarmasin in 1603 while the English visited the country for the first time in 1615, both in an effort to procure pepper (Van Dijk 1862; Young 1991).

The increasing demand for pepper in this period, however, put the sultanate in a precarious position as the pepper trade raised a conflict of interest between the involved parties. The expansion of the sultan's pepper plantation to the upstream area that encroached on the land of the Dayaks, the natives of Kalimantan, led the Dayaks to attack the capital. Mataram, the major Javanese kingdom, attacked the city after it became known that the sultanate, having gained considerable profits from the pepper trade, intended to repudiate its subordination as a vassal of Mataram. In the meantime, a disagreement broke out between the sultan—who would rather have had a free market—and the European traders—who desired a pepper monopoly. The dispute culminated in the sultan destroying the Dutch mission when, in 1606, it pursued an exclusive commercial policy under the Vereenigde Oost-Indische Compagnie (VOC; Dutch East Indies

12 In his report Beeckman named the Kuin River as the Tatas Small River (see Beeckman 1718, 59).

Company) envoy Gillis Michielszoon. The Dutch retaliated in 1612, severely damaging the capital. This precarious situation, and in consideration of Banjar's vulnerability as being easily accessed by the enemy from the sea and upstream, influenced the reigning Sultan Marhum Panembahan (r. 1595–1620) to move the court and the port with its *syahbandar* upstream (Van Dijk 1862; Ras 1968).

The shifting of the capital to Kayu Tangi and subsequently to Martapura caused Banjar to lose its port-polity functions. Nevertheless, Banjar still played a role in the sultanate's commercial activities. With its strategic geographic location, Banjarmasin became a secondary port city, a crucial peripheral node in the sultanate's pepper trade network. Foreign vessels moored at the Banjar port, which at that time was known as Old Banjar, dispatched smaller ships to the capital to conduct trade negotiations with the sultan. Banjarmasin as a subsidiary city, therefore, supported the economic function of the upriver port-polity.

The insecure conditions and the transfer of the court and main port from Banjar are presumed to have had an impact on this area. The port became a landmark of the town, replaced the court, and consequently turned it into an urban center. The locals and the sojourning traders dwelled on the riverside or stayed in their vessels close to the port on the Kuin River. However, not all of the population lived in Old Banjar, since some moved to Tatas, an area nearby the confluence of the Kuin and Martapura rivers on the eastern side of Banjarmasin (Beeckman 1718; Fong 1969) (see Figure 4.2). Among the population, a significant number of the Javanese migrants who had fled from the Mataram aggression formed the community of Kampung Jawa in Tatas (Valentijn 1724; Fong 1969).

Although Banjar was no longer the capital of the sultanate, the town maintained its character as a Southeast Asian port city but without performing its function as a political center. Trading, both wholesale and retail, was centered in this town and furthermore influenced the social organization which was evidenced in the important role of the *orang kaya*, who acted as the middleman between foreign traders and the sultan (Beeckman 1718; Reid 1980). The heterogeneous society persisted; the population lived in clusters based on their origins and ethnicity beside the Kuin and Martapura rivers. The predominant swamp and riverine vegetation gave the town an urban-village look, while the waterways provided primary access and a means of transportation that supported the population's activities. It appears that the impact on Banjar of the political economy of the pepper trade turned it into a secondary port city, which changed the spatial and social structure of its urban patterns.

The Imprints of the Foreign Traders: Banjarmasin as the Sultanate's Primary Port (1663–1787)

By 1663 the sultanate's principal port shifted from Martapura to downstream Old Banjar and then around the 1690s to Tatas. The first transfer was probably due to several reasons. The first was the power sharing between two of the sultanate's princes who then divided their authority into Martapura and Old Banjar. The second was the longer traveling time needed to reach the port of Martapura. From J.J. De Roy's account we learn that the sultanate's port-of-trade existed in Tatas since at least 1691 and was more than likely there before then (De Roy 1700, 67). A pragmatic reason for the port's relocation to Tatas could be that the Martapura River could be sailed by large ships, while the Kuin River was shallow, especially in the Antasan Kuin area. Thus the Kuin was unsuitable for large vessels (Müller 1857), such as the Chinese junks, which arrived in Banjarmasin in increasing numbers for the pepper trade as the sultanate had become recognized by foreign traders as a pepper producer. To reach Tatas, ships from the Java Sea sailed upstream through the Barito River and approached the port through the mouth of the Martapura. Since the Martapura followed a meandering course, the sultanate cut off one of the bends near Tatas to shorten the route.[13] Following the appointment of Tatas as the main port-of-trade, it is believed that the Old Banjar port turned into a local harbor.

As the principal port, in the 1690s Tatas evolved into a new urban center. Here, the sultanate government was represented by the *syahbandar* and the toll-house.[14] The importance of trade for the sultanate was marked by the existence of the sultan's residence at Tatas, where he sometimes stayed in order to control his territory or to negotiate with foreign traders, the presence of 700 men posted there to act as security, and a guard post on the riverbank of the Martapura adjacent to the port to regulate the arriving and departing ships (De Roy 1700; Valentijn 1724; Müller 1857; Gais 1922; Noorlander 1935). These features all indicate that Tatas was highly significant, being critical in terms of generating revenue for the sultan. This led to Tatas to be recognized as the city of Banjarmasin while the role of Old Banjar gradually receded.

The flourishing commerce in eighteenth-century Banjarmasin was characterized by competition among the dominant traders, i.e., the English with the East India Company (EIC), the Chinese junks, and the Dutch with

13 Map interpretation NA 4.VEL 371, 1740.

14 ANRI Marginalia, 1707.

the VOC.[15] In this period, the EIC and the VOC attempted to obtain exclusive pepper trading privileges from the sultan. Meanwhile, Chinese merchants projected Banjarmasin as their new source of pepper (Fong 1969). These interests caused intense rivalry among the foreign traders which is deemed to have generated influential processes on changes to the landscape and left material and spatial footprints on this river city.

The English Fortification

The early stages of eighteenth-century trade in Banjarmasin was marked by English domination of the pepper commerce (Gais 1922; Suntharalingam 1963; Fong 1969). Considering Banjar's strategic location, the EIC planned to establish a commercial post in Banjar that served as the intermediate port for transshipment trading with the Chinese junks. Therefore, they proposed the establishment of a fort, a proposal which was initially rejected by Sultan Bagus Kusuma but was eventually granted after the sultan was defeated in the first Anglo-Banjarese War of 1701. The on-land fort construction on the north side of the Kuin River mouth in Old Banjar was an extensive project that employed 200 Bugis laborers and required wood imported from Java and bricks from China.[16] The completion of this work took a significant period due to several obstacles, including the swampy condition of the site, the climate and high precipitation, the lack of skilled laborers and rice to feed them, and, to a certain extent, a lack of wood. Whilst the fort was being built, commercial activities in Tatas went into decline since the wholesale trade shifted to the EIC port. However, after the second Anglo-Banjarese War in 1707, which demolished the fort, trading resumed and Tatas regained its importance. In the years following this, the sultan refused the EIC permission to build a fort but instead allowed the opening of a factory in Tatas (Gais 1922; Fong 1969).[17]

The Chinese Influence

Chinese traders played a significant role in Banjar's economic sector. The traders who visited Kalimantan came from the coastal cities of East Kwangtung, Fujian, and Canton on the south coast of mainland China. To

15 Other foreign traders—such as the Portuguese, Spanish, Danish, and French—rarely visited Banjarmasin (see Knapen 2001, 69). It is therefore presumed that they did not have a significant impact on the landscape. Although the Portuguese traded intensively during the last quarter of the seventeenth century (Fong 1969, 39), and might have had an influence on the landscape, this is not further analyzed in this study.

16 Interpretation of ANRI Marginalia, 1707; CL G258:2/17, 1703; and Beeckman (1718).

17 ANRI Marginalia, 1707.

expand their trading network, some of them sojourned and others settled in Banjarmasin. Their ventures increased in this town mainly from the early seventeenth century with between four and fifteen junks annually sailing to Banjarmasin (Chang 1954; Fong 1970; Lockard 2013). The junks were loaded with various commodities from China, such as silk, earthenware, copper, tea, and other goods required by the local people of Banjarmasin. In exchange, the Chinese bought Bornean products, such as dragon's blood, canes, birds' nests, gold, blue glass beads, but mainly pepper (Gais 1922; Suntharalingam 1963). Compared to European traders, the Chinese offered a higher price for pepper, which resulted in the local agents preferring to sell their products to the Chinese merchants. In this way, Chinese traders were on the one hand accepted by the sultan and his people and on the other became a significant competitor to the European traders in the pepper trade (Gais 1922; Suntharalingam 1963; Fong 1970; Sulandjari 1991; Young 1991).

The Chinese were granted permission to establish a settlement on the eastern river bank of the Martapura River across from the port in Tatas. Just like the locals, they extended their compound on the levee, but many lived in floating houses on the river (Müller 1857; Sulandjari 1991). After the second Anglo-Banjarese War, the number of Chinese settlers allegedly increased from 80 families in 1707 to 200 in 1715.[18] They engaged in wholesale and retail trading by using their floating houses as shops. The shops were tastefully decorated, which made this *kampung* unique. While the number of junks allowed to enter the port sometimes was restricted to only one junk a year due to the VOC prohibition since it held the exclusive pepper trade, their arrival was highly anticipated. The royal family and the people celebrated their arrival with festivities by organizing a floating fair adjacent to the Chinese *kampung*. This event not only created a social space on the river but also articulated the economic function of the Martapura River and Tatas. Presumably because of the significant presence of the Chinese junks, the Martapura River was also known as the China River in the early eighteenth century (Beeckman 1718; Van der Kemp 1898; Gais 1922).

The Dutch Establishment

Contact between the sultan and the VOC in the early eighteenth century was occasional rather than persistent. Whenever the EIC participated in trade, the VOC was absent in Banjarmasin. After the exile of the EIC in 1707, the VOC re-established their enterprise in this city (ANRI 1965). The VOC occupied a factory in Tatas and held the trade monopoly for

18 ANRI Marginalia, 1707.

several periods as compensation for providing military assistance to the sultan during the confrontation with the Dayaks and the Bugis. Although the sultan allowed the Dutch to build a stone fortress as stipulated in the 1756 contract, they continued to use the existing warehouse in Tatas (ANRI 1965).[19] Nevertheless, this modest factory, which was erected on elevated ground with an area of about 700 square meters surrounded by the swamp, was burned down in 1761 by order of the then reigning Sultan Panembahan Kaharudin Halillulah (r. 1761–1801), in order to expel the VOC from his territory. However, the sultan's threat did not discourage the Dutch from retaining their privilege in the pepper trade and consequently they rebuilt the factory. The basic structure and rural look of the establishment of the VOC in Tatas and the prevailing riverine life and swamp environment in this town is thought to have prevailed until the 1780s (Noorlander 1935).

The above discussion reveals that the presence of foreign traders generated the social space through the compound formation along the rivers. The dwellers amplified the pluralism in the city, while the river-front settlements intensified the river port city milieu. Whereas the Europeans applied their homeland traditions to build structures on land, the Chinese adapted to the local conditions and used the vernacular architecture (see Figure 4.5).[20]

Swamp Exploitation: Banjarmasin as the Primary Dutch Port (1787–1860)

In 1787, the sultan and the VOC agreed on a new contract in which the sultan ceded his hegemony and consented to share the territory with the VOC. The VOC land encompassed the coastal area, including Tatas and the port that surrounded the sultan's land. The sultan and the VOC territories were delineated by the Kuin River and *Antasan* ("canal") Kuin. From then on, the VOC exercised direct governance over its land, and, particularly, the control over trade. To strengthen its authority, the Dutch improved its trading post in Tatas and turned it into a fortress that supported the military, trade, and settlement functions (see Figure 4.4). The fort was constructed on the raised bank and was enclosed by a wooden palisade inside which were placed the military barracks, a house of residence, and

19 See ANRI (1965, 67: Contract 1756, article 27).

20 Before 1850 Chinese migrants elsewhere were known for their willingness to adapt to the local culture and social life, enabling them to build a social network with the local society without difficulty (see Lockard 2013).

Figure 4.4 Reconstructed map of the landscape of Banjarmasin in the 1830s that shows Dutch intervention on the landscape mainly through the construction of road, canals (*antasan*), and swamp-rice cultivation (Adapted from Vera D. Damayanti, 2019)

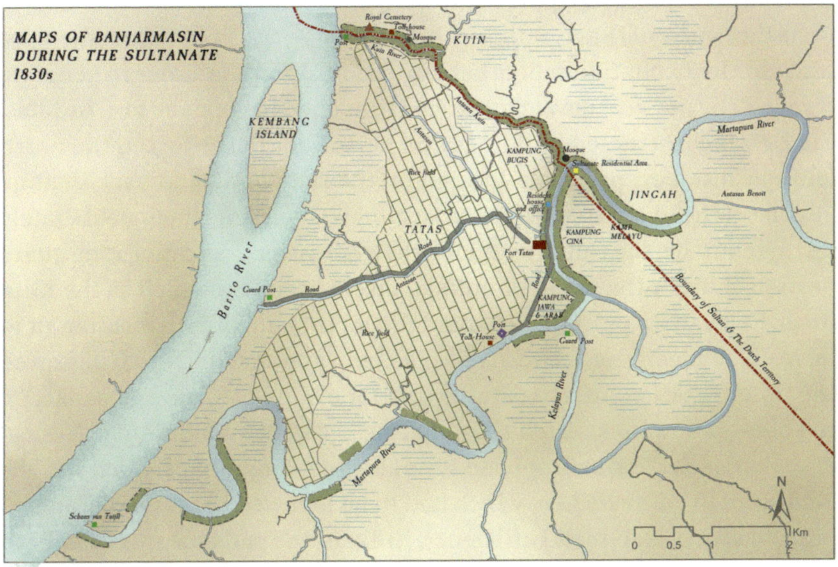

a storage house, while a toll-house was placed adjacent to the port (Müller 1857; Van der Kamp 1898; ANRI 1965).

During the Dutch East Indies administration of Governor-General Herman Willem Daendels (1808–1811), the Dutch withdrew from the Banjarmasin post since their commercial endeavor there had become unprofitable. They left the city on 9 September 1809 and shortly after, upon the sultan's request, the EIC took over from the Dutch as stated in the contract of 1 October 1812. The English occupation was brief due to the signing of the London Treaty between England and the Netherlands in 1814 which compelled the restoration of the Dutch East Indies colony. One event that had a significant impact on the landscape during the EIC occupation was probably the removal of the royal families from Tatas in order to comply with the 1812 contract that caused the sultan to transfer his residence to Kampung Mesa, on the other side of Tatas, north of the Chinese compound (ANRI 1965).[21]

In accordance with the new agreement of 1 January 1817, the Dutch, authorized by the Dutch East Indies government under the *Zuider-en*

21 Further evidence is required to validate the proposition.

Ooster-afdeeling van Borneo ("the South and East Part of Borneo Residency"), returned to occupy the port of Tatas and the former VOC land with Banjarmasin—referred to as Tatas—as the capital.[22] This designation promoted the recognition of Tatas, to be known as Banjarmasin, while Old Banjar was gradually perceived as a part of the Banjarese Kampung Kuin.

As the course of Dutch authority in the archipelago shifted from the spice trade to the exploitation of other natural resources in order to generate revenue, the physical development had to expand beyond the fort. In Tatas, the Dutch constructed buildings and ports near to the fort in the north and south side respectively, located on the river bank and back swamp. Furthermore, in anticipation of smuggling and piracy, which deliberately disrupted trade, the Dutch extended their defensive area by locating guard posts at the mouths of the prominent rivers near Tatas, such as the Kuin River, Kelayan River, and Martapura River.[23] Meanwhile, in an attempt to increase connectivity between one place and another, the Dutch improved the transportation infrastructure with, for example, roads and canals on the back swamp to shorten the ship line.[24] Besides conducting physical developments, the Dutch applied several regulations, one of which was the obligation to cultivate rice, which had to be followed by all inhabitants in order to achieve food self-sufficiency. In Tatas, the outcome was evidenced by the expansion of the paddy fields in the 1820s that in turn shaped the urban-village scenery of this town (Knapen 2001) (see Figure 4.4).

The Dutch occupation apparently had a profound influence on the transformation of the landscape of Banjar. Infrastructural and agricultural development on the swamp changed the structure of the urban fabric which also caused the loss of the swamp forest and the peat layer. The political and economic interests between the sultan and the Dutch, however, had consequences with regard to the reduction of the sultan's authority and territorial division. Related to this, waterways, i.e., the Kuin River and canal, became the physical border between these two jurisdictions, while

22 The sultan more than likely agreed to the contract because he needed military protection from enemy and pirate attacks, which continued until the 1810s (see Van der Kemp 1898, 151; Knapen 2001, 241). From the contract it could be inferred that the north of the Kuin River, formerly known as Old Banjar, was considered as Kampung Kuin, with no reference to the existence of Old Banjar.

23 NA NL-HaNA, Osten, 2.21.205.45, 1818, 1822.

24 From Müller's map (1857) which depicts Banjarmasin in 1836, there were three *antasan* ("artificial channels"), i.e., Antasan Besar, which connected Fort Tatas and the Barito River, Antasan Benoit, which cut off the crossing of a meandering channel of the Martapura River, and another *antasan* that connected the Kuin River and Fort Tatas.

Figure 4.5 Depiction of the landscape of Banjarmasin in 1843 which shows the Chinese *kampung* on the left bank and Fort Tatas on the right bank of the Martapura River (Reprinted with permission from KITLV [37A169])

from an economic perspective, the waterways played a role in controlling access for trade.

This analysis of the transformation of the area explains the character of modern Banjarmasin's landscape in the context of urban spatial configuration, landscape elements, and people's activities. The transfer of power in politics and the economy from the sultan to the Dutch caused the transformation of Tatas, formerly the sultanate's urban center on the west bank of the Martapura River. The structure was changed from a traditional to a modern urban typology with regard to the spatial arrangement and material features. The history of this power shift can be seen to date in the urban spatial structure of Banjarmasin city, in which the main urban centers located on the riverside of Martapura River—such as the Banjarmasin municipality offices, the main markets of Pasar Sudimampir, Pasar Lama, and Pasar Baru, and the old port—were formerly the main areas of the Dutch settlement. The existence of the sultanate as part of the urban heritage remains on the Kuin riverside which has become a Banjarese *kampung* where traces of the sultanate can still be seen in the preserved Royal Cemetery of the sultanate and the Suriansyah Mosque. The settlement along the rivers with pole houses and floating houses, which previously prevailed in a great number along the main rivers, still exist, although the number is significantly lower in the case of the floating houses (*lanting*) due to the lack of materials required for the construction of this vernacular form of housing. The attachment of local people to the river has been maintained, which is reflected in the river-based daily activities and the use of the waterways for transportation. The new floating market on the east bank of Martapura River, which has been revitalized and has become one of the city's main tourist attractions, marks the revived tradition of the river being used for economic activities.

Although the physical structure of Banjarmasin as a river city may be sustainable, it is the intrinsic value of river-based culture that was threatened by the possibility of its disappearance. Over the past decade the municipal government has focused its attention on the river conditions, which it had previously neglected. However, this river revitalization somehow does not integrate the cultural values of the river. Considering the fact that the physical and cultural landscape of Banjarmasin is characterized by the river, the urban development of this city requires the incorporation of the cultural aspects of the river, particularly, to preserve the identity of Banjarmasin as the thousand river city.

Conclusion

The stages of transformation of the Banjarmasin landscape during the sultanate period confirm that the political-economic factor played an important role in shaping the townscape. The political-economic power of the sultanate in the South Kalimantan region could not be separated from the main rivers of Barito and Martapura that had a central role in controlling the trade in inland commodities and imported goods, a typical pattern found in the chiefdoms and sultanates in Sumatra, Borneo, and Malaysia that had great rivers (Sutherland 2007). The transformation of the cityscape involved key processes such as warfare and contact with foreign traders, political and economic aspects that significantly influenced the metamorphosis of Banjarmasin from a port-polity to a port-of-trade controlled by the Dutch. In this transformation, the tidal swamp and rivers considerably affected the structure of the urban fabric that formed the river city of Banjarmasin.

The change in Banjarmasin's landscape showed a reciprocal human-landscape interaction that characterized the cityscape. The fundamental process that initiated the formation of Banjarmasin was the population's need of space for performing their basic activities, for which they adapted to the conditions of the swamp and fluvial terrain. In the regional context, Banjarmasin's location on the south coast, close to the mouth of the Barito River, affected the political and economic structure of Banjar society. From an economic point of view, and considering that the swamp soil was unsuitable for agriculture, trade as the primary source of income was the main occupation. Its geographical position, that is, its closeness to the mouth of the Barito River, also enabled Banjarmasin to maintain control over the upstream-downstream trade network. From a political perspective, this connection further influenced the structure of state formation of the sultanate with regard to becoming a port-polity where in this political system the upstream areas became the subordinates of the sultan who was seated downstream. However, this geographical position characterized by easy access somehow became a drawback from the political perspective (Boomgaard 2007). It made the city vulnerable to outsider attack and caused warfare driven mainly by economic interests such as those witnessed in Banjarmasin.

On the city scale, orientation to the river was imperative because it was the only means of communication and transportation. Consequently, proximity to the river was indispensable. Moreover, a location close to the confluence of the two main rivers was strategic since it resulted in control over the mobility and transfer of goods. Therefore, the sultanate selected sites near the junction of the Kuin and Barito Rivers for its political and economic centers, and in the late 1690s moved to the confluence of the Martapura and Kuin rivers.

On the other hand, the political-economic activities of Banjarmasin had an impact on the waterways which was manifested through several interventions. The first is that of the use of space on the river and on the riverside to accommodate the city's needs affected the intrinsic functional value of the rivers, which was derived from political and economic functions as well as residential and socio-cultural functions. The port, floating market, and transportation line are prominent aspects of the spatial use of the river derived from its economic functions. The use of the Kuin River as a territorial boundary between the sultan and the Dutch in the eighteenth century indicates that the river had a political purpose. Another intervention that has left a physical change on the landscape, which persists today, is the construction of canals. For instance, the sultanate extended the Kuin River streams through a canal to link the Barito and Martapura rivers and cut off the bend of the Martapura River to shorten the shipping route in order to facilitate trade activity. Meanwhile, the Dutch built several canals in the back swamp at Tatas which connected their fort to the Kuin and

Barito rivers for security and military defense. Therefore, the functional spaces and the construction of hydrological structures contributed to the urban form of Banjarmasin as a river city, which also indicates an extensive relationship between the city and the river.

Over time the Banjarmasin landscape changed. The rivers and canals still exist as the city's deep context and as the major urban landscape feature. Moreover, human-riverscape interaction can still be observed in city life. Rivers, as part of the urban ecosystem, keep providing services to the locals. Concerning the classification of ecosystem services (Gómez-Baggethun et al. 2013), the waterways in Banjarmasin today play roles as a provisioning water source, regulating urban hydrology, and providing intangible services related to the economy, transportation, recreation, and identity. All are interrelated and shape the cultural landscape of Banjarmasin that to a great extent conveys "water culture." Boomgaard (2007) has asserted that "water culture" will change from one period to another and may shift either positively or negatively. In Banjarmasin, despite the local people's dependency on the river decreasing as compared to during the sultanate era, the "water culture" is still embedded in the city through its physical landscape and the perceived intangible values. The "water culture" has inevitably become the urban heritage of Banjarmasin. It is, therefore, imperative that the municipality consider the "water culture" and its ecosystem services when considering the sustainable urban development of Banjarmasin.

References

Archival Documents

Arsip Nasional Republik Indonesia, Jakarta (ANRI), Marginalia to the Daily Journals 1659–1807, Item een rapport van den ondercoopman Jan de Wolf wegens desselfs wedervaren tot Banjar, 18 July 1707, file 2531, folio 499–518.

Caird Library, London (CL), G258:2/17, map, A large draught of the south part of Borneo by Sami Thornton at the signe of England Scotland and Ireland in the Minories London, produced 1703.

Leiden University Libraries, KITLV 37A169, "Südlicher Theil von Banjermassing," Heinrich von Gaffron, 1843.

Nationaal Archief, Den Haag (NA) NA 4.VEL 371, map, Kaart van de Rivier van Bandjermasin, Anoniem, ca. 1740.

NA NL-HaNA, Osten, 2.21.205.45, inv.nr. 9, Rapporten betreffende de forten en posten te Bandjermasin, 1818, 1822.

Published Sources

Ahyat, Ita S. 2012. *Kesultanan Banjarmasin Pada Abad ke-19: Ekspansi Pemerintah Hindia-Belanda di Kalimantan*. Tangerang: Serat Alam Media.

ANRI. 1965. *Surat-surat Perdjandjian antara Kesultanan Bandjarmasin dengan Pemerintahan2 VOC, Bataafse Reppublik, Inggeris dan Hindia Belanda 1635–1860*. Jakarta: ANRI Kompartemen Perhubungan dengan Rakjat.

Beeckman, Daniel. 1718. *A Voyage to and from the Island of Borneo, in the East-Indies*. London: printed for T. Warner and J. Batley.

Bock, Carl. 1882. *The Head Hunters of Borneo: A Narrative of Travel Up the Mahakkam and Down the Barito; Also, Journeyings in Sumatra*. London: S. Low, Marston, Searle, & Rivington.

Boomgaard, Peter. 2007. "A State of Flux Water as a Deadly and a Life-Giving Force in Southeast Asia." In *A World of Water: Rain, Rivers, and Seas in Southeast Asian Histories*, edited by Peter Boomgaard, 1–24. Leiden: KITLV Press.

Bronson, Benneth. 1977. "Exchange at the Upstream and Downstream Ends: Notes Toward a Functional Model of the Coastal State in Southeast Asia." In *Economic Exchange and Social Interaction in Southeast Asia: Perspectives from Prehistory, History, and Ethnography*, edited by Karl L. Hutterer, 39–52. Ann Arbor, MI: University of Michigan Press.

Chafee, John. 2008. "At the Intersection of Empire and World Trade: The Chinese Port City of Quanzhou (Zaitun), Eleventh-Fifteenth Centuries." In *Secondary Cities and Urban Networking in the Indian Ocean Realm, c. 1400–1800*, edited by Kenneth R. Hall, 99–122. Lanham, MD: Lexington Books.

Chang, Tsuen-Kung. 1954. *Historical Geography of Chinese Settlement in the Malay Archipelago*. PhD Dissertation. University of Nebraska.

Damayanti, Vera D. 2019. "Identifikasi Struktur dan Perubahan Lanskap Banjarmasin di masa Kesultanan (1526–1860)." *Jurnal Lansekap Indonesia*, 5(2): 249–259.

De Roy, Jacob J. 1700. *Voyagie gedaan door Jacob Janssen de Roy na Borneo en Atchin, in het jaar 1691*. Batavia.

Fong, Chin Yoon. 1970. "The Chinese in Banjarmasin during the First Half of the Eighteenth Century." *Journal of the Historical Society, University of Malaya*, 8: 66–72.

Fong, Goh Yoon. 1969. *Trade and Politics in Banjarmasin, 1700–1747*. PhD dissertation. University of London.

Gais, Johannes W. 1922. *The Early Relations of England and Borneo to 1805*. PhD dissertation. University of Berne.

Gómez-Baggethun, Erik, Åsa Gren, David N. Barton, Johannes Langemeyer, Timon McPhearson, Patrick O'Farrell, Erik Andersson, Zoé Hamstead, and Peleg Kremer. 2013. "Urban Ecosystem Services." In *Urbanization, Biodiversity and Ecosystem*

Services: Challenges and Opportunities, edited by Thomas Elmqvist et al., 176–251. Cham: Springer. doi: 10.1007/978-94-007-7088-1_11.

Irhamna, Umi Sriwahyuni, and Irhamsyah Safari. 2001. *Masjid Sultan Suriansyah*. Banjarmasin: Panitia Pemugaran dan Pengembangan Masjid Bersejarah Sultan Suriansyah.

Kathirithamby-Wells, Jeyamalar. 1990. "Introduction: An Overview." In *The Southeast Asian Port and Polity: Rise and Demise*, edited by J. Kathirithamby-Wells and John Villiers, 1–16. Singapore: Singapore University Press.

Knapen, Han. 2001. *Forests of Fortune? The Environmental History of Southeast Borneo, 1600–1880*. PhD dissertation. University of Nijmegen.

Lockard, Craig A. 2013. "Chinese Migration and Settlement in Southeast Asia before 1850s: Making Fields from the Sea." *History Compass*, 11(2): 765–781. doi: 10.111/hic3.12079.

McGee, Terence G. 1969. *The Southeast Asian City: A Social Geography of the Primate Cities of Southeast Asia*. London: Bell and Sons.

Müller, Salomon, 1857. *Reizen en onderzoekingen in den Indischen Archipel, gedaan op last der Nederlandsche Indische regering, tusschen de jaren 1828 en 1836*. Amsterdam: F. Muller.

Noorlander, Johannes C. 1935. *Bandjarmasin en de Compagnie in de Tweede Helft der 18de Eeuw*. PhD dissertation. University of Leiden.

Ras, Johannes J. 1968. *Hikajat Bandjar: A Study in Malay Historiography*. PhD dissertation. University of Leiden.

Reid, Anthony J.S. 1980. "The Structure of Cities in Southeast Asia, Fifteenth to Seventeenth Centuries." *Journal of Southeast Asian Studies*, 11(2): 235–236.

Reid, Anthony J.S. 1988. *Southeast Asia in the Age of Commerce, 1450–1680. Vol. 1: The Lands below the Winds*. New Haven, CT: Yale University Press.

Reid, Anthony J.S. 1993. *Southeast Asia in the Age of Commerce, 1450–1680. Vol. 2: Expansion and Crisis*. New Haven, CT: Yale University Press.

Sulandjari. 1991. *Politik dan Perdagangan Lada di Kesultanan Banjarmasin (1747–1787)*. MA Thesis. University of Indonesia.

Sumawinata, Basuki. 1998. "Sediments of the Lower Barito Basin in South Kalimantan: Fossil Pollen Composition." *Tonan Ajia Kenkyu*, 36(3): 293–316.

Suntharalingam, Ramanathan. 1963. "The British in Banjarmasin: An Abortive Attempt at Settlement 1700–1707." *Journal of Southeast Asian History*, 4(2): 33–50. doi: 10.1017/S0217781100002 79.

Sutherland, Heather. 2007. "Geography as Destiny? The Role of Water in Southeast Asian History." In *A World of Water: Rain, Rivers, and Seas in Southeast Asian Histories*, edited by Peter Boomgaard, 27–70. Leiden: KITLV Press.

Tagliacozzo, Eric. 2007. "An Urban Ocean Notes on the Historical Evolution of Coastal Cities in Greater Southeast Asia." *Journal of Urban History*, 33(6): 911–932. doi: 10.1177/0096144207304035.

Valentijn, François. 1724. *Oud en nieuw Oost-Indiën Deel 3, Derde Boek*. Dordrecht: Joannes van Braam.

Van der Kemp, Pieter H. 1898. "Het Afbreken van onze betrekkingen met Bandjermasin onder Daendels en de hersteling van het Nederlandsch gezag aldaar op den 1e januari 1817." *Bijdragen tot de Taal-, Land- en Volkenkunde van Nederlandsch-Indië*, 49(1): 1–168.

Van Dijk, Ludovicus C.D. 1862. *Neêrland's Vroegste Betrekkingen Met Borneo, Den Solo-Archipel, Cambodja, Siam En Cochin-China ... Met Eene Levensschets En Inleiding Van Mr. G.w. Vreede*. Amsterdam: J.H. Scheltema.

Van Wijk, Christiaan L. 1951. *Soil Survey of the Tidal Swamps of South Borneo in Connection with the Agricultural Possibilities*. Contributions of the General Agricultural Research Station, Bogor, Indonesia, No. 123. Bogor: Ministry of Agriculture.

Young, Robert J. 1991. "The English Factory at Banjar on the Island of Borneo, 1699–1707." *Indonesia Circle. School of Oriental and African Studies Newsletter*, 19(55): 3–16.

Biography

Vera Dian Damayanti is a lecturer at the Department of Landscape Architecture, IPB University, Indonesia and is currently working on her PhD at the Centre for Landscape Studies, University of Groningen, the Netherlands. Her research focuses on the cultural biography of the river city landscape of Banjarmasin.

5 Through the Passage and across the Worlds

The Bridge-and-Temple Complex in the Popular Processions in a "Rurban" Town in Jiangnan Region[1]

Xiaomei Zhao

Abstract

In the sixteenth century, in China's Jiangnan Region, market towns along the canals started urbanizing. However, because the urbanization process was not completed, these towns were classified as rural settlements. In this chapter I refer to these in-between-the-rural-and-the-urban towns as "rurban," a status implied in popular religious rituals such as processions through and beneath the bridges in front of temples. Taking one such procession through a bridge-and-temple complex during the Notable Shi's birthday celebration in Fengjing Town as an example, I argue that the complex has a specific religious signification and social function, and that the continuity of such signification cannot be achieved without suitably adapting the function.

Keywords: bridge-and-temple complex, popular procession, rurban town, Jiangnan Region, Fengjing Town

Introduction

The region of Jiangnan, south of the lower reaches of the Yangtze River, occupies the core area of the Yangtze River Delta Economic Zone. This region, comprising the Shanghai metropolis as well as parts of Jiangsu

1 This chapter has been published in Chinese in the journal *Zhuangshi* in 2019. [赵晓梅. (2019). "江南市镇迎神赛会中的桥庙空间研究——以枫泾北镇施王寿诞香汛为例." 装饰, 8: 86–89.]

Padawangi, Rita, Paul Rabé and Adrian Perkasa, *River Cities in Asia*.
Taylor & Francis Group, 2022
DOI: 10.5117/9789463721851_CH05

and Zhejiang provinces, has been the most economically and culturally developed area in China since the twelfth century. The urbanizing process in the agricultural villages of this region began with rural market places in the sixteenth century (Fu 1964; Zhu 1997; Atsutoshi 2008[2001]) and, as transport by water was the main traffic route for rural commerce, generated market towns along the canals. However, their urbanization was incomplete. The word "rurban" indicates this status of unfinished urbanization. These towns fell somewhere between urban cities and rural villages; their residents considered themselves neither citizens of cities nor peasants of villages. Nonetheless, in the administrative framework that prevailed during the Ming and Qing Dynasties they belonged to the rural system. These rurban towns, located along the canals, represented a unique type of river settlement in the history of China. And their popular rituals demonstrated this status between the urban and the rural (Atsutoshi 1991b, 1995, 2008[2001]; Wang 2010). For example, worship of the City God [城隍 *chenghuang*] was only officially allowed in the cities, but temples to the City God were built in the towns to mimic cities and to govern the rural villages that fell under the unofficial township territory.

These towns have become the religious centers within their market networks. Many temples were established along the canals, often situated by a bridge. Processions were carried out during specific events for the deities, joined by pilgrims [香客 *xiangke*] from the rural villages. Although research has been carried out on the social organization of the popular processions in the market towns (such as Yu 2014), the use of the space is rarely touched upon. The complex of bridges and temples has religious implications in the procession, and thus serves as a lens through which to understand the social function of the popular procession, and social relations within the township territory. In this chapter I elaborate on the origins and evolution of the complex in the rurban towns, especially taking the bridge as the "passage" (see the concept in Gennep 1960[1909]) to the temple in the processions. Guided by ritual specialists, pilgrims use and perceive this passage differently according to their social roles in the processions, which are socially important events for rurban society. Therefore, the spatial representation of the complex not only contains religious significance but also incorporates participants' social relations.

Firstly, I propose that the three dimensions of the passage in the bridge-and-temple complex represent a spatial framework in the religious centers of rurban towns. Secondly, the processions in Fengjing Town are used as illustrations to explain the spatial representation. In the past, social groups undertook various performances in the processions, made different use of the complex, and perceived the multidimensional passages differently. I then present the changes that have emerged in both the procession and the social relations among participants in the current event. I argue that this very complex is essential in the conception of religious importance and social function of the popular procession in these towns, as evidenced by people moving along the passages and putting on their own performances. Without proper spatial practice, the social and cultural representation of the complex can hardly be achieved in the current rurban society.

Popular Religion in the "Rurban" Towns in Jiangnan Region

The Yangtze River is the second most important river—after the Yellow River—in China's history. It is the longest river in Asia, originating in the Tibetan Plateau in the west of China and running all the way across the country to its eastern coast. Jiangnan Region, located at the river's eastern end, is a flatland of canals and lakes, and has been one of China's most developed areas since the Southern Song Dynasty moved its capital city to the region in the twelfth century. The soil was fertile and good for growing rice and other commercial crops, enabling the area to support a large population. Eventually, due to population pressure, land became scarce and by the sixteenth century commercial markets had gradually become the main industry in the rural settlements behind farming and fishing (Atsutoshi 1988, 2007). This led to the rise of the market towns. Generally speaking, towns in China originated by providing periodic markets for their nearby villages in the pre-modern peasant marketing network (Skinner 1964).[2] The villages became the *xiangjiao* [乡脚] for the towns, i.e., participants in and affiliated to the market town (Fei 1984).

From the sixteenth to the eighteenth century the towns underwent urbanization (Fan 1998), but the trend was never officially recognized by the authorities. In the traditional Chinese sense, only a city was entitled to the walled residence of the prefecture and county (Friedmann 2007). In contrast, a town simply had waterway borders. They were not considered rural either, as most townspeople were merchant gentry, traders, and artisans, rather than peasants and fishermen (Fan 1990). Furthermore, there was no *yamen*

2 The period in which towns emerged depended on the region. In the case of Jiangnan, it was from the sixteenth to the eighteenth century; much later in other regions.

[衙门], the government office which held jurisdiction in such a town, but only a branch office for tax collection and postal service during the Qing Dynasty (Zhang 2008; Yu 2010). Nevertheless, the unofficial administrative division did not always coincide with the market network, as some of the towns were situated right at the border between counties [县 *xian*] or even prefectures [府 *fu*]—for example, the towns of Fengjing [枫泾], Nanxiang [南翔], and Liantang [练塘], whose affiliated rural communities could be found on both sides of the borders.

The towns were what I would call "rurban"—between the urban and the rural. This indicates the social position of the towns and their residents within the Chinese system of urban and rural settlements (Xiong 2019). A town served as the connecting point between the urban and rural in the fields of marketing and rituals. It was the social and cultural center of the rural society, only secondary to the city to which it was affiliated. Thus the events that took place in a town involved not only its own townspeople, but also the peasants from surrounding villages, and sometimes the urban citizens of the city to which it was affiliated.

In the Jiangnan Region of wetlands and lakes, a town was always established along a canal, as water connected the urban and rural settlements and thus facilitated trading. However, water also brought flood and catastrophe, the prevention of which was provided by rituals and offerings. Similar rituals for and along waterways have been practiced widely in Asia and beyond since ancient times, such as in the cases of Angkor (Hang 2015), India (see Chapter 3 of this volume), and Egypt (Fazio et al. 2008). Even though C.K. Yang's (1967[1961]) concept of diffused religion is still open to discussion, the social functions of popular religion in China have been accepted by many scholars, especially in their role in the governing (Feuchtwang 1991; Zhao 1998) and social structures (Hsu 1973). The popular rituals enacted in the towns were different to those performed in the rural villages and in the urban cities, thus implying their rurban status. Temples to the City God were commonly found in the towns even though, at the beginning of the fourteenth century, they were only allowed in the cities as part of state rituals (Atsutoshi 1991b, 1995, 2008[2001]). The establishment of a township's temple to the City God revealed the ambiguous identity of the townspeople in the urbanizing process. The deity of a certain territory was authorized as a Heavenly Minister with the same ranking as an official in the imperial governing system on earth. Ideally, a town's City God was subordinate to the God of its affiliating county, and superior to the Earth Deities [土地公 *tudi gong*] of its affiliated villages (Atsutoshi 1991a, 2008 [2001]; Wang 2010). This system mirrors exactly the governing system. The temple territory [

庙界 *miaojie*], within which people would go on pilgrimage to the temple (Wang 2010), was in general coincident with the administrative division, except in the case of bordering towns. In the latter situation, there were dual temples, just as there were dual governmental offices under different counties.

There were many other local deities in the region, the worship of whom was always connected to daily life. Many deities were in charge of water, because fishing was a main source of living; while other deities were in charge of agriculture and health (Gu 1994; Fan 2006). The deities were sometimes given different rankings, as were officials in the government, as in Jinze Town, for example.[3] The City God usually worked as the local minister, to whom all other deities were required to pay homage in the popular processions during *xiangxun* [香汛], the specific event for a certain deity (Yu 2014; Weller 2015).

Processions are practiced in many religions all over the world. In China, they could be found in villages, towns, and cities by the water and in the mountains. The processions enabled the local deity to announce their territory, in other words, they served to solidify a community under a certain spiritual authority. In the case of the rurban towns, the townspeople and the peasants from their affiliated villages were connected and united in such an event, even if they played different roles in the procession.

The religious rituals also mimicked governmental actions. One of the important rituals was *jie-tianxiang* [解天饷], literally meaning "dispatching heavenly tax." The aim was to collect spirit-money (tax) for the central township temple(s) from its affiliated temples and from households in the town and its affiliated villages. The spirit-money was then sent to Heaven, which is to say, the Eastern Peak [东岳大帝 *dongyue dadi*] or Jade Emperor [玉皇大帝 *yuhuang dadi*] (Atsutoshi 2008[2001]; Wang 2010; Yu 2014; Tao and Goossaert 2015). As the local City God was in charge of the tax collection in his territory, his image was compelled to watch over the ritual, in front of the images of the heavenly gods.

The deities were brought around the town to collect the "tax" and to give blessings to their patrons, including both the townspeople and the peasants from the town's affiliated villages. The incense societies [香社 *xiangshe*], which were groups of organized pilgrims from the rural communities, walked behind the images of the deities. During the procession, the ritual masters (法师 *fashi*, i.e., the Taoist priests in Jiangnan Region) performed

3 There were many temples in Jinze and the main gods of the temples constituted the Heavenly government, according to interviews with locals in July 2017.

rites for the local patrons among the townspeople. The procession started from the central temple, navigated the bridges along the main roads and canals—essentially crossing the sacred and the profane worlds—and finally returned to the same temple for the final burning sacrifice. The rituals performed during the popular processions and their organization in Jiangnan Region have been largely covered in the research by Wang (2010), Yu (2014), and Tao and Goossaert (2015). The use of the space in the procession by different social groups, however, has been rarely studied. The bridge-and-temple complex is one of the most important ritual spaces whose religious meanings in the processions are essential for understanding the social history of the rurban towns, as a type of river "cities." Trade and traffic along canals brought about the rise of the towns and facilitated their urbanisation. The towns have been established on the wetlands by the canals, and their social life is always connected to water, such as in the popular processions, in which the land and water spaces have been used and linked. Such rurban towns and their rituals cannot be separated from their water resource, as we can see from diverse cultural practices of water heritage around the world (Hein et al. 2020).

Methods

There has been a mountain of research on popular religion in China in the fields of history, anthropology, and religious studies. Scholars have noted the social function of religious rituals in strengthening a person's attachment to a territorial community. Furthermore, the concept of temple territory has been established, in which the space and the territory are critical for the social relations of the participants in the religious events in the region (Wang 2010).

The bridge-and-temple complex accommodates significant rituals and has diverse meanings in the popular processions. This research uncovers the triple passages of the complex based on reviews of literature on the bridge rituals in China. The possible routes of the processions in Fengjing are reproduced utilizing mainly maps and descriptions in township gazettes [镇志 *zhenzhi*] from the late nineteenth and early twentieth centuries, when the town prospered economically and religiously. Unfortunately, there are very few records. The authors were usually local Confucian intellectuals who considered the popular religion unorthodox and did not write much about the rituals (Zhao 2003; Yu 2014). To supplement the limited written notes, fieldwork was conducted. Structured interviews were carried out,

Figure 5.1 Locations of rurban towns in metropolitan Shanghai (Reproduced by Xiaomei Zhao)

in July 2017, with dozens of locals in the streets or in their houses in several towns, including Xinchang [新场], Fengjing, Jinze, and Liantang, which is currently part of the Shanghai metropolis (Figure 5.1). In each town, a group interview with local scholars and governmental officials was also held.[4]

4 The group interviews were conducted as part of the "Research on cultural ecology of famous historical towns in Shanghai" program at the Land and Cultural Resource Research Center, Fudan University.

In September 2017 I also conducted participant observation of the rituals during a *xiangxun* in Fengjing Town.

From the historical and ethnographical study, the bridge-and-temple complex can be understood as a social space (Lefebvre 1992[1938]), in which different types and levels of social groups have collaborated, competed, and even fought with each other. They have also produced and reproduced their social relations by using this ritual space, with changes made throughout history to adapt to their various social needs.

Bridge and Temple as a Religious Complex

Bridges are largely constructed in the settlements along rivers and canals for transport purposes. The water beneath the bridge is considered as Yin [阴] and needs to be suppressed by talismanic objects to avoid floods and other disasters (Wang 2012). Initially, a temple may have been affiliated to a bridge (Tu 2015), such as a shrine of a local deity in the middle of or at one end of a bridge.[5] The shrine is often emphasized by a magnificent roof indicating the vertical passage to Yang [阳]. The shrine is the center of the universe, whose roof, sometimes together with a caisson ceiling and a special pattern on the floor, implies the "*universalis columna*" (see the definition in Eliade 1987[1956]), a passage from the netherworld of the water, through this world of human beings to Heaven. This is the first passage of the bridge-and-temple (-shrine) complex. This vertical passage is not very common in Jiangnan Region. The Wanan Bridge [万安桥] in Jinze Town [金泽镇] is an exceptional case. There used to be a pavilion built on the bridge whose upper floor accommodated a shrine of Guanyu [关羽], a popular deity all over China. In an interview, a local recalled her childhood as a little girl playing and running beneath the shrine,[6] stressing the vertical spatial levels of the gods and the human.

The passage turns from vertical to horizontal when the shrine is enlarged into a temple and moves onto the bank at the end of the bridge.[7] A bridge, therefore, works as an essential passage in rituals, with a strong

5 The complex of a shrine at a bridge is common in southern China. Outside China, the LaiViễnKiều (known as the Japanese Covered Bridge) in Hoi An, Viet Nam may serve as an example.

6 The interview was carried out by the author in July 2017.

7 The bridge is generally older than the temples built at its ends. But I disagree with the idea proposed by Li (2017) that the temple was built by the bridge for the convenience of transportation.

suggestion of being a "Bridge of Sighs" between This World and Beyond (Yang 1967[1961]), and can be found in many rituals in southern China (Qin 1992; Zhou 2011[1993]). Moreover, throughout China, walking across the bridges [走桥 *zou qiao*] is a common custom during festivals, to ward off illness and catastrophes (Chen 2017; Zhou 1998). In some rituals a bridge can be "implied" to serve a transitional function—for example, a paper bridge is made for a funeral (Zhu and Tan 1996). This passageway is found in non-religious architectural complexes as well. In many grand palaces in China, a bridge is built over a canal or a pool in front of the first main hall after the gate. Walking through the bridge highlights the formal beginning of the socially elevated hierarchical realm.[8] In the rurban towns of Jiangnan Region, a bridge is commonly situated right in front of a temple. The bridge primarily serves as an entrance to the temple from the other side of the water, and at the same time signifies a transitional space to the sacred world from the profane one. In some cases, there might even be two temples, one at each end, making the bridge the passage between two different sacred worlds. The bridges are important nodal points along the route of popular processions during a *xiangxun* (see the maps of the procession routes in Yu 2014) and serve as transitional passageways between the different worlds defined by the temples.

In Jiangnan Region, another type of procession takes place on the water along the main canals (usually called market canal [市河 *shi he*] on the sides of which the markets are located). These processions are conducted in the evening and take place in addition to the on-land daytime processions along the main roads (Zhang 2010). A third passageway is thus established. The waterway was mainly used for transport in the region, with most pilgrims traveling by boat. Similar to the processions on land, the images of the deities are seated on board a boat which journeys through the town and passes underneath all the monumental bridges (Yu 2014). The individual pilgrims and the incense societies then follow the deities in their own boats.

All three passages co-exist in the bridge-and-temple complex in Jiangnan Region. The complex has been produced and reproduced throughout history for its practical function and religious rites. The usage of the complex by different social groups represents the social relations in the township and beyond, linking the on-land society to its water resource.

8 This idea was inspired by a personal communication with Tao Jin, an architect specializing in Taoism.

The Popular Procession in Fengjing Town

A rurban town is often situated on the border of a waterway transport hub. Fengjing is one such border town. According to the township gazetteers (Xu et al. 2005[1891]; Cheng 2005[1910]), a local market was accommodated in its current location since the Song Dynasty period (1237–1240), and was given the status of "town" during the Yuan Dynasty period (1341–1370). During the Ming Dynasty period (1426–1435), its southern and northern parts fell under Jiashan County of Jiaxing Prefecture and Huating County of Songjiang Prefecture respectively. Lou County split from Huating and became the superior of the northern towns during the Qing Dynasty period (1638–1661).

Dual administration of the town led to dual residence for deities. There used to be a major Taoist and a major Buddhist temple in either part, as well as temples to the City God, Notable Shi [施王 *shi wang*], Guanyu, and Duke Yan [晏公 *yan gong*]. However, there was only one independent residence for the Eastern Peak in the southern town, whose location was quite close to the border (Figure 5.2). According to the historical records, the processions in Fengjing had been big and bustling since the mid-eighteenth century. There were processions on the birthdays of the City Gods in spring and of the northern Notable Shi in autumn (Government of Fengwei Town of Jinshan County 1993; Guan 2007). The celebration took place either before or after the intensive farming season so that the peasants would be available to take part and may also have served as an occasion to pray for or to celebrate a good harvest.

The Notable Shi is a local deity and presides over snakes, rain, tide, and health (Fan 2006), all of which are crucial for local peasants and fishermen. His temple was one of two central temples in the northern town, the other being the temple of the City God. Since the late Qing Dynasty period the procession on his birthday has been the biggest (Guan 2007), in which his images traveled both on land and on water. The incense societies from nearby villages followed the images of the deities and put on performances in the procession (Zhang 2010). The deity went and paid respect to the local minister, at the temple of the City God of the northern town, and the procession started from there. Then the deity traveled through the northern town[9] along the main roads and the market canals. The procession may also have roamed a little into the southern town where the Eastern Peak is located. Guilds constructed temporary sheds for the ritual masters to

9 The temple territory of the northern Notable Shi would likely have covered only the northern town, as there was another temple to the Notable Shi in the southern town. The procession would therefore have only been conducted within the northern town.

Figure 5.2 Historic map of Fengjing Town (Reproduced by Xiaomei Zhao)

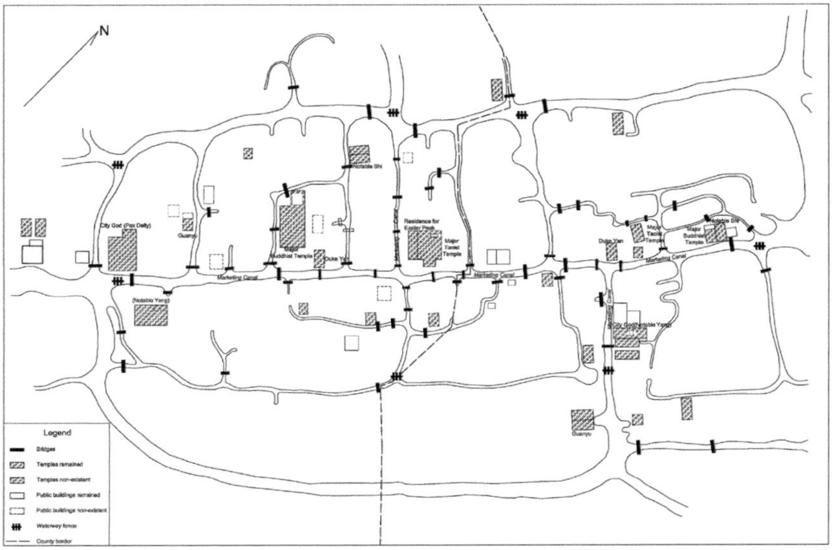

conduct the rites. Finally, a troupe invited by the committee of the pilgrimage association [香会 *xianghui*], who were the coordinators of this *xiangxun*, would have staged a dramatic performance for the deities' amusement (Atsutoshi 2008[2001]; Fan 2006; Yu 2014).

The routes of the procession were designated by the ritual masters and the pilgrimage association. There is little historical record of the exact routes. Nevertheless, the primary priest that currently works as the coordinator of the *xiangxun* informed me of the principles for a route plan.[10] The route must be a closed journey around the entire (northern) town, and the paths must be broad enough for the large processional troupe. Additionally, several key stops in the procession are named in records (Guan 2007) and thus it has been possible to locate them on a map. Figure 5.3 reproduces the possible routes, which must have passed both across many bridges on land and beneath them on canals.

The ritual masters led the processions and conducted certain rites along the routes (Guan 2007; Yu 2014). The stops designated for the performance of these rites would most likely have been at important temples or at the temporary sheds constructed by the guilds. They would also most probably have been next to or very close to the bridges, which is where most temples

10 I interviewed him on the day before the birthday celebration in September 2017.

Figure 5.3 Possible routes of the popular procession on the birthday of the northern Notable Shi in Fengjing Town (Reproduced by Xiaomei Zhao)

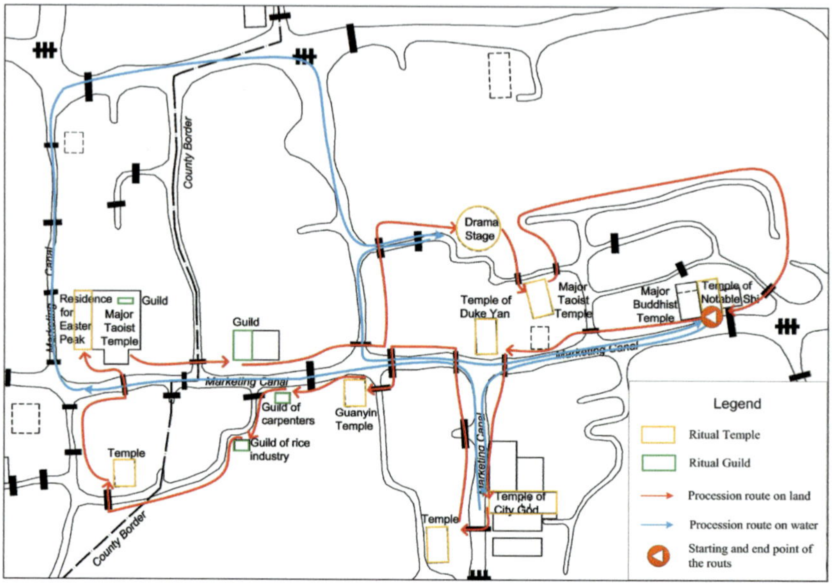

and guild halls were situated along the canals. Moreover, only at bridge ends would the space be big enough for such rites to be conducted, because the paths in the town were generally not wide enough. The rites performed by the masters were primarily for the patrons (Tao and Goossaert 2015), while performances by incense societies were conducted more for their own sake rather than for an audience (Yu 2014). The masters worked as agents connecting Heaven and this world by guiding the troupes through all the transitional passages of bridges from one stop to another. The association, on the other hand, collected the money and coordinated different groups financially.

The pilgrims and patrons used the space according to their different social relations with the deity, or with the town as the religious center. The rural folk had their own space at the Temple of the Earth Deity, which was hierarchically lower than that of the township deity (Atsutoshi 2008[2001]). The communities formed incense societies, which followed the processions through the passages. The townspeople, on the other hand, were divided by their occupations into guilds. The townsfolk guilds may not have journeyed all over the town but stayed at their own spots. These people probably came from outside the administrative division, because the entire town is a marketplace. There is no record of any procession for the Notable Shi in the southern part of the town; villagers from the southern administrative division also take part in the current event.[11]

Current Changes in the "Procession"

From the 1950s to the 1980s the popular religions were forbidden, but since the 1990s they have been revived in the region (Li 2017). Unfortunately, during the last decades, the southern part of Fengjing has been almost totally demolished to make way for new construction, with no temples remaining. But in the northern part, in recent years, the temples to the City God and the Notable Shi have been rebuilt. In addition, the two *xiangxuns* on the deities' birthdays have been preserved.

The restored popular rituals are still conducted by the Taoist priests working in the temples, whose social roles are no longer limited to being ritual specialists. All the priests received official training on Taoist rituals from a government-funded program[12] and their position in the temples is similar to that of civil servants in the governmental offices. They cannot organize the event according to their own wishes, as all religious activities must be sanctioned by the Office of Religious Affairs Administration. The procession of the images of the deities has been forbidden in metropolitan Shanghai. Nevertheless, the incense societies still hold a procession through the town and stage their own performance, as I observed in Fengjing on the occasion of the Northern Shi's birthday in 2017.

In the seventeenth and eighteenth centuries the event originally took three days, but increased to seven days by the end of the nineteenth century (Zhang 2010). It was then reduced to only one day during the twentieth century but extended to three days in recent decades.[13] The priests are in charge of all the rituals in the temple. They move the small images of the Notable Shi and his wife from the upper room of the wing hall on the first day of the event, and set them at the gate hall, facing the main hall where the unmovable statue of the same deity is located (Figure 5.4). The deities receive homage from the pilgrims in the courtyard, rather than from all over town as they would have done during the procession in the past.

11 According to fieldwork conducted during the *xiangxun* in September 2017.

12 According to an interview with the primary priest in Fengjing in September 2017.

13 Interview with the primary priest in Fengjing in September 2017. The shrinking was due to both economic decline and government decree, while the revival is evidence of the change of policy on popular religion.

Figure 5.4 Movable images of the deities are moved to the courtyard where pilgrims pay homage to them in the northern temple of Notable Shi in Fengjing (Photo by Xiaomei Zhao, 2017)

However, because the rites require more priests to perform them than the ones stationed at the temple, several priests from other temples in Shanghai come to help with the rites on all three days.

Thousands of pilgrims take part in the three-day event. Most people, mainly elderly women from nearby rural villages, attend with their friends and family. They come every day and "pay their tax" willingly according to what they can afford. There are now only two incense societies from nearby villages. One of them carries out a procession through the town (without the images of the deities), while the other, the smaller group, performs in the temple courtyard. The processioning group comes from a village in Jiashan County in Zhejiang Province,[14] less than 20 kilometers and only a 15-minute drive away. The current administrative division does not cut off the original religious sphere.

Besides the traditional pilgrims, several newly emerged groups come from the town or even from metropolitan Shanghai. These groups contribute much more financially than the rural pilgrims and are grand patrons for

14 The village used to be within the administrative division of the southern town of Fengjing.

this event.[15] Many rites are reserved primarily for them, as they stand right in front of the deities and receive the deities' blessing ahead of the other pilgrims. These groups comprise newly wealthy pilgrims; they come together through individual connections such as with colleagues and neighbors and are organized by spirit-mediums [灵媒 *lingmei*] (Tao and Goossaert 2015). The latter play the coordinating role formerly played by the pilgrimage association.

Today, there is no role for the pilgrimage association as there was in the past. The major coordinators are the Taoist priests and the spirit-mediums, whose social networks have established the current temple communities.[16] The pilgrimage is no longer an event that reflects the communal construction within the traditionally multi-level temple territory. The pilgrims are more diverse than territorial communities or guilds. The locals now take part in *xiangxuns* for different deities rather than just the ones in the central temples of their own towns.[17] The modern efficient transport network may have worked as a driver of this change. From the social function perspective, the event seems more individualized on the one hand, while on the other, the state has even more control of the activities because the priests as the main coordinators are practically civil functionaries.

The utility and directivity of the space has also changed. The movable and unmovable images of the Notable Shi are set face to face across the temple courtyard. The pilgrims offer their incense to both. The rites conducted by the priests take place primarily in the main hall. Most rites are to the unmovable image (Figure 5.5), rather than to the movable images that used to be the major icons that sat at the front of the procession. The priests conduct rites to the movable image in the courtyard only at the beginning and at the end of the event. The final burning of spirit-money takes place in the courtyard in front of the unmovable image after the movable image has already been carried back to the upper floor (Figure 5.6). Nevertheless, following tradition, the dramatic performance, invited by the priests, is set up on a contemporary stage facing the movable image.

The implication of the tax collection for Heaven has almost disappeared. The Eastern Peak residence had been abandoned by the end of the nineteenth

15 According to interviews with the Taoist priests, the temple managers, and the pilgrims themselves in September 2017.

16 According to the primary priest of this temple. My personal observation verifies his statement: pilgrims and spirit-mediums seem to know him and they all conversed with him during the event.

17 Pilgrims from the rural villages near Fengjing told me about some of the different religious events beyond the administrative division of Fengjing.

Figure 5.5 A rite conducted by the Taoist priests to the unmovable image of Notable Shi in the main hall (Photo by Xiaomei Zhao)

Figure 5.6 Burning the spirit-money in the courtyard at the end of the three-day event (Photo by Xiaomei Zhao)

century (Xu et al. 2005[1891]; Cheng 2005[1910]). Even the image of the Jade Emperor, who still resides in the wing hall of this temple, does not show up at any of the rites conducted during the three-day event, including the final burning. The rites have consequently lost part of their original meanings due to the changes in spatial use.

Hardly any use is made of the bridge-and-temple complex. The incense societies still conduct processions, but the trip is much shorter, only connecting the Notable Shi and the City God (who is temporarily enshrined in a rebuilt Buddhist temple to the north of the town, because his own temple is under construction), rather than covering the entire territory. The only group that goes on procession consists of two dozen middle-aged men and women, carrying flags representing either a family or a societal group. The group only performs on the third day. They come into the temple, procession into the main hall, and then form a circle in the courtyard. Afterwards they walk all the way to the shrine of the City God, offering donations to the deities, and then return to the temple of the Notable Shi, once again forming a circle in the courtyard. This is the end of the procession. The original spatial experiences through all the transitional spaces of the bridges have been lost. They do not pass through any of the old bridges anymore, because the procession takes place mostly outside the old town. There is no master in the troupe to conduct any rites along the route, and there is no audience waiting for the rites to be conducted.

The waterway procession has also ceased to exist. There is no longer any transport by water and the canals have been made narrow or converted into land for construction. Thus no pilgrims arrive by water. Moreover, the location of the temples has been changed: they have been rebuilt along the main roads rather than situated right by the canals. And although pilgrims continue to hold processions to certain specific sacred sites, like the older generation used to, the procession has lost the significance of passing through all the passageways, with no masters to guide the procession or to conduct the proper rites. All these factors have led not only to the meaningful spatial resonance of the bridge-and-temple complex becoming inaccessible, but also to society keeping away from the water.

Discussions

The emergence of rurban towns in Jiangnan Region in the sixteenth century announced the beginning of urbanization in China (Fu 1964; Zhu 1997; Atsutoshi 2008[2001]). The towns first developed as the marketplace for the local rural area, and then became the religious and administrative center. At that time the popular religion and its rituals in these towns implied their rurban position in the imperial governing system (Li 2014). Thus the temple territory may have coincided with the market network and the administrative division. However, this territory would also have been constantly changing with the economic development in the region and competition among the towns (Wang 2010). The towns established their own temples to the City God, whose rituals included the popular procession and originated from the rural areas (Yu 2014). The townscape, with all the main canals and roads, was much more delicate than the rural landscape in the villages. And the bridges were the intersections between the waterways and on-land routes.

The bridge-and-temple complex—especially the covered bridge—is common in South China (Dai 2005). Temples are usually affiliated to the bridges. In the towns of Jiangnan, as the temple gained more and more followers, the complex evolved into something different. They have been enlarged and, as a result, detached from the bridge and situated on the banks of the waterways. The bridge has thus become a horizontal passageway to the temple. Consequently, the bridge and the temples are relatively independent. The vertical passageway of the bridge is no longer essential in Jiangnan towns and today, hardly any important rites are performed on the bridge itself. However, the bridges provide significant transitional passageways during the processions, either through the bridge in the on-land route or underneath the bridge on the canal.

Several social groups used to take part in this religious practice and the town's merchant gentry and artisans used to play a vital role in the procession. They established associations and became the main patrons and coordinators of the event. Intellectuals worked as the negotiators between the government officials and the pilgrims if any conflict arose (Fei 1980; Yu 2014), though they may not have taken part directly in these activities. The masters were the spiritual guides, conducting rites along the route, and were followed by organized groups and individual pilgrims from rural villages. As we have seen from the case of Fengjing, participants' social relations and roles have been completely altered. Trained by the government, the priests are the main organizers, and also serve as the masters guiding the pilgrims. However, due to official control the procession of the deity has not been restored. The incense society continues to hold the procession, but by choosing a different route that does not include crossing bridges or passing underneath them, the group does not follow tradition. More diverse pilgrims, such as urban citizens from metropolitan Shanghai, take part, though these pilgrims have personal connections to the priests and the spirit-mediums, and are not connected by residing within the township territory.

Changes have also occurred in the current administrative framework of urban and rural settlements in China. Many market towns have been given the right of jurisdiction and have gained an officially recognized government status below the county level. Accordingly, the townspeople are urban citizens, even though they do not consider themselves comparable with city dwellers. Fengjing is one of these administrative towns. But its territory has shrunk compared to when the villages were affiliated to it in the past. To a certain extent, this has led to less participation by rural incense societies, even though a group from its former territory continues with their traditional practices. These changes are immutable, but the establishment of new social relations are being encouraged by both the continuation of traditional religious rituals and the adaptation of cultural practices.

In general, the connection with water has been reduced significantly. The traffic on land dominates in daily life as well as in the rituals. This facilitates pilgrims coming from afar, so that people from beyond the temple's former territory can take part. These pilgrims are even able to visit several temples within the same day. However, the absence of water also causes the disappearance of the religious implications of passage in the bridge-and-temple complex. The demolition of temple buildings and the reduction of the procession, as well as the lack of rites conducted by the ritual masters, all contribute to the current situation. Participation in the rituals has also become individualized rather than collective due to the breaking down of traditional social relations. Locals still follow the custom of "walking through the bridges and worshipping the deities in the temples" [走桥拜庙 *zouqiao baimiao*], however, they no longer fully understand its meaning in the spatial practice. In an interview with a rural man performing in the incense society, he said he does not know why they hold the procession. He is simply following what his father and his grandfather used to do.

The spatial representation of the bridge-and-temple complex lies not only in the transitional rites but also in the social history of the rurban towns by the canal. However, this has not been given enough attention by urban planners and architects. These professionals never consider the bridge-and-temple complex as a spatial combination in religious rituals to reproduce the social relations within the township and beyond, but are merely concerned about the bridge as an element of the townscape and the temple as a cultural place. This has led to the misplacement of newly built temples and bridges, which aggravates the inconceivability of the religious significance of the complex. Further research on the history of the rituals related to water is needed, as well as on the continuation of traditional practices. This needs to be done following an adaptive approach to re-establish the connection between the people and the water, and between the urban and the rural.

Acknowledgments

I owe deep gratitude to Mr. Tao Jin, who has given me a lot of help in writing this paper. His ideas on Taoist rites have inspired me greatly. I am very grateful to Prof. Torgrim Guttormsen, who read my draft and provided helpful suggestions. I thank all the symposium participants in Surabaya whose questions and comments have expanded my thoughts. It is my good fortune to work with the team at Fudan University and to take part in the research projects on the conservation of the water towns in Shanghai. Last but not the least, I want to express my appreciation to the symposium coordinators and the book editors, Dr. Paul Rabé, Mr. Adrian Perkasa, and Dr. Rita Padawangi, without whom such good communication among the scholars from different fields would not have been possible.

References

Atsutoshi, Hamashima. 1988. “Land Development and Traveling Trade of the Landowners in Jiangnan in Middle Ming Dynasty.” *Guangdong Social Sciences*, 2: 84–86. [滨岛敦俊. “明代中叶江南土地开发和地主的客商活动.”广东社会科学.]

Atsutoshi, Hamashima. 1991a. “Research on the City God in Jiangnan Region in Ming and Qing Dynasties.” *The Journal of Chinese Social and Economic History*, 1: 39–48, 108. [滨岛敦俊. “明清江南城隍考——商品经济的发达与农民信仰.” 中国社会经济史研究.]

Atsutoshi, Hamashima. 1991b. “Research on the City God in Early Ming Dynasty.” *Social Scientist*, 6: 22–30, 21. [滨岛敦俊. “明初城隍考.”社会科学家.]

Atsutoshi, Hamashima. 1995. “Research on the Revolution of City God by Zhu Yuanzhang.” *Collected Papers of History Studies*, 4: 7–15. [滨岛敦俊. “朱元璋政权城隍改制考.”史学集刊.]

Atsutoshi, Hamashima. 2007. “Settlements, Communities and Peasants Relations in Jiangnan Region.” *Society*, 27(3): 189–20. [滨岛敦俊. “江南的聚落、社区与农民共同关系.” 社会.]

Atsutoshi, Hamashima. 2008[2001]. *Rural Societies and Popular Religions in Jiangnan Region in Ming and Qing Dynasties*. Xiamen: Xiamen University Press. [滨岛敦俊. 明清江南农村社会与民间信仰. 厦门: 厦门大学出版社.]

Chen, Enwei. 2017. "The Origin and Evolution of 'Zoubaibing'." *Folklore Studies*, 2: 42–50. [陈恩维. "'走百病'民俗的渊源与流变." 民俗研究.]

Cheng, Jianshan. 2005[1910]. *Supplementation of Fengjing Chorography*. Shanghai: Shanghai Academy of Social Sciences Press. [程兼善. 续修枫泾小志. 上海: 上海社会科学院出版社.]

Dai, Zhijian. 2005. *China Covered Bridges*. Fuzhou: Fujian People's Publishing House. [戴志坚. 中国廊桥. 福州: 福建人民出版社.]

Eliade, Mircea. 1987[1956]. *The Sacred and the Profane: The Nature of Religion*. New York, NY: Harper & Row.

Fan, Jinmin. 1998. *Commerce Development in Jiangnan in Ming and Qing Dynasties*. Nanjing: Nanjing University Press. [范金民. 明清江南商业的发展. 南京: 南京大学出版社.]

Fan, Shuzhi. 1990. *Research on the Towns in Jiangnan in Ming and Qing Dynasties*. Shanghai: Fudan University Press. [樊树志. 明清江南市镇探微. 上海: 复旦大学出版社.]

Fan, Ying. 2006. *Research on Popular Religion in Shanghai*. Shanghai: Shanghai People's Publishing House. [范荧. 上海民间信仰研究. 上海: 上海人民出版社.]

Fazio, Michael, Marian Moffett, and Lawrence Wodehouse. 2008. *A World History of Architecture*. London: McGraw-Hill.

Fei, Hsiao-tung. 1980. *China's Gentry: Essays on Rural-Urban Relations*. Chicago, IL: University of Chicago Press.

Fei, Hsiao-tung. 1984. "Small Towns, Big Problems (1): Diverse Towns in Wujiang." *Outlook Weekly*, 2: 18–20. [费孝通. "小城镇 大问题（之一）——各具特色的吴江小城镇." 瞭望周刊.]

Feuchtwang, Stephan. 1991. *The Imperial Metaphor: Popular Religion in China*. London: Routledge.

Friedmann, John. 2007. "Reflections on Place and Place-Making in the Cities of China." *International Journal of Urban and Regional Research*, 31(6): 257–279.

Fu, Yiling. 1964. "Research on the Economics in the Towns in Jiangnan Region in Ming and Qing Dynasties." *History Teaching*, 5: 9–13. [傅衣凌. "明清时代江南市镇经济分析." 历史教学.]

Gennep, Arnold van. 1960[1909]. *The Rites of Passage*. Chicago, IL: University of Chicago Press.

Government of Fengwei Town of Jinshan County, Shanghai (ed.). 1993. *Chorography of Fengwei Town*. Shanghai: Shanghai Popular Science Press. [上海市金山县枫围乡人民政府 (编). 枫围乡志. 上海: 上海科学普及出版社.]

Gu, Xijia. 1994. "Historical Research on the Popular Religion in Jiangnan Region." *Religious Studies*, 4: 59–69. [顾希佳. "江南地区民间信仰沿革考." 宗教学研究.]

Guan, Fulin. 2007. "Temple Fairs and Peasant Market in Jinshan." In *Jinshan Book in Shanghai Historical Documents*, edited by CPPCC Committee of Jinshan,

Shanghai; Committee of Historical Documents of Shanghai CPPCC, 227–231. Shanghai: Xinhua Bookstore. [管甫霖. "漫话金山庙会农市." In政协上海市金山区委员会, 政协上海市委员会文史资料委员会 (编) 上海文史资料选辑金山卷. 上海: 新华书店.]

Hang, Peou. 2015. "Water and Heritage in Angkor, Cambodia: The Monuments, the Ancient Hydraulic Network and Their Recent Rehabilitation." In *Water and Heritage: Material, Conceptual and Spiritual Connections*, edited by Willem Willems and Henk van Schaik, 71–86. Leiden: Sidestone Press.

Hein, Carola, Henk van Schaik, Diederik Six, Tino Mager, Jan Kolen, Maurits Ertsen, Steffen Nijhuis, and Gerdy Verschuure-Stuip. 2020. "Introduction: Connecting Water and Heritage for the Future." In *Adaptive Strategies for Water Heritage: Past, Present and Future*, edited by Carola Hein, 1–18. Cham: Springer.

Hsu, Chia-ming. 1973. "Territorial Organization of Hoklorized Hakka in the Changhua Plain." *Bulletin of the Institute of Ethnology Academia Sinica*, 36: 165–190. [许嘉明. "彰化平原福佬客的地域組織." 中央研究院民族學研究所集刊.]

Lefebvre, Henri. 1992[1938]. *The Production of Space*. Hoboken, NJ: Wiley-Blackwell.

Li, Tian-gang. 2014. "The Relationship between the Local and the National in the Sacrifice System of Jiangnan: A Case Study on Jinze Town, Suzhou Prefecture and Songjiang Prefecture." *Journal of East China Normal University* (*Philosophical Social Science Edition*), 4: 69–79. [李天纲. "江南镇乡祭祀体系中的地方与国家——以上海金泽镇及苏、松二府为例."华东师范大学学报 (哲学社会科学版).]

Li, Tian-gang. 2017. *Jinze: The Origin of Popular Sacrifice in Jiangnan*. Shanghai: Sdxjoint Publishing Company. [李天纲. 金泽：江南民间祭祀探源. 上海: 生活·读书·新知三联书店.]

Qin, Shaoshan. 1992. "The Bridge Customs in Southern China." *Journal of Zhongnan College of Nationalities* (*Humanities and Social Science*), 5: 85–87, 101. [覃韶山. "中国南方民族桥俗浅探." 中南民族学院学报 (哲学社会科学版).]

Skinner, William. 1964. "Marketing and Social Structure in Rural China, I." *Journal of Asia Studies*, 24(10): 195–228.

Tao, Jin, and Vincent Goossaert. 2015. "Daoism and Local Society in Suzhou." In *Religions and Public Life in Jiangnan Region*, edited by Robert Weller, 86–112. Shanghai: Shanghai People's Publishing House. [陶金, 高万桑. "道教与苏州地方社会." In魏乐博 (编). 江南地区的宗教与公共生活. 上海: 上海人民出版社.]

Tu, Shiping. 2015. "Classification and Interpretation of Chinese Ancient Water Guardians." *Journal of Zhejiang University of Water Resource and Electric Power*, 27(3): 1–6. [涂师平. "我国古代镇水神物的分类和文化解读." 浙江水利水电学院学报.]

Wang, Jian. 2010. *At Stake: A Research on the Popular Religion in Suzhou and Song in Jiangnan Region since Ming and Qing Dynasties*. Shanghai: Shanghai People's

Publishing House. [王健. 利害相关: 明清以来江南苏松地区民间信仰研究. 上海, 上海人民出版社.]

Wang, Peijun. 2012. "Water Guardians and Traditional Talismanic Customs in China." *Journal of Hohai University* (*Philosophy and Social Science Edition*), 14(2): 53–57, 92. [王培君.镇水兽与中国传统镇水习俗. 河海大学学报 (哲学社会科学版).]

Weller, Robert. 2015. "Introduction: Religion and Social Changes in Great Jiangnan Region." In *Religions and Public Life in Jiangnan Region*, edited by Robert Weller, 1–17. Shanghai: Shanghai People's Publishing House. [魏乐博. "导言: 大江南地区的宗教与社会变迁"In魏乐博 (编). 江南地区的宗教与公共生活. 上海: 上海人民出版社.]

Xiong, Wansheng. 2019. "'Urban-Rural Society': A New Concept for Understanding China's Urban-Rural Relation." *Beijing Cultural Review*, 1: 46–53, 142. [熊万胜. "城乡社会: 理解中国城乡关系的新概念."文化纵横.]

Xu, Guangyong, Shixiong Ye, and Yun Fei. 2005[1891]. *Recompilation of Fengjing Chorography*. Shanghai: Shanghai Academy of Social Sciences Press. [许光墉, 叶世熊, 费沄. 重辑枫泾小志. 上海: 上海社会科学院出版社.]

Yang, C.K. 1967[1961]. *Religion in Chinese Society: A Study of Contemporary Social Functions of Religion and Some of Their Historical Factors*. Berkeley, CA: University of California Press.

Yu, Shuangyuan. 2010. "Administrative Division and Collaboration of the Towns on Borders in Qing Dynasty: A Case Study of Zhangliantang Town in Jiangnan." *Journal of Bohai University*, 2: 73–77. [于双远.清代小跨界市镇的政区分隶与合作——以清代江南章练塘镇为例. 渤海大学学报.]

Yu, Zhejun. 2014. *Deities and City Dwellers: Research on Popular Processions in Shanghai during the Republican Era*. Shanghai: Shanghai Joint Publishing Company. [郁喆隽. 神明与市民: 民国时期上海地区迎神赛会研究. 上海: 三联书店.]

Zhang, Haiying. 2008. "The Administrative Modes of Towns South of the Yangtze River during the Ming and Qing Dynasties." *Academic Monthly*, 40(7): 130–139. [张海英. "明清江南市镇的行政管理." 学术月刊.]

Zhang, Weiling. 2010. "Temple Fair." In *History of Fengjing*, 230–233. Shanghai: Shanghai Literature and Art Press. [张玮玲.枫泾庙会In枫泾史话. 上海: 上海文艺出版社.]

Zhao, Shiyu. 1998. "Interactions between State Sacrifice and Popular Religion: A Case Study of Journal of Ding in National Capital and Eastern Peak Temple in Ming and Qing Dynasties." *Journal of Beijing Normal University* (*Social Science Edition*), 6: 18–26. [赵世瑜. "国家正祀与民间信仰的互动——以明清京师的"顶"与东岳庙为个案."北京师范大学学报(社会科学版).]

Zhao, Shiyu. 2003. "Religious Activities, Leisure Life and Female Subculture of the Women since Ming and Qing Dynasties." In *Popular Religion and Social Space*, edited by Zhenman Zheng and Chunsheng Chen, 148–182. Fuzhou: Fujian People's Publishing House. [赵世瑜. "明清以来妇女的宗教活动、闲暇生活与女性亚文化."In郑振满, 陈春声 (编) 民间信仰与社会空间. 福州: 福建人民出版社.]

Zhou, Xing. 1998. *Boundary and Symbol: Bridges and Folk Customs*. Shanghai: Shanghai Literature and Art Press. [周星.境界与象征:桥和民俗. 上海: 上海文艺出版社.]

Zhou, Xing. 2011. "Han Customs of Bridge." In *The Logic of Country Life*, 175–197. Beijing: Peking University Press. [周星. 汉民族的桥俗文化In乡土生活的逻辑. 北京: 北京大学出版社.]

Zhu, Jianming, and Jingde Tan. 1996. *Bridge Rituals and Culture in the Peasant Families in Laogang Town, Nanhui County, Shanghai*. Taipei: Zhenghezhen Foundation. [朱建明; 谈敬德.上海南汇县老港乡农家渡桥仪式与桥文化. 台北: 施合郑基金会.]

Zhu, Xiaotian. 1997. *Modern Changes of Society of Jiangnan Towns*. Beijing: China Business Press. [朱小田. 江南乡镇社会的近代转型. 北京: 中国商业出版社.]

Biography

Dr. Xiaomei Zhao is an associate professor on heritage studies and architectural history at the Department of Cultural Heritage and Museology, Fudan University, Shanghai. She is trained in the fields of archaeology and architectural history. Her research interests include vernacular architecture, rural settlement, critical heritage studies, and the conservation of built heritage.

6 Challenges to an Iconic River-City

Contesting Communities, Threatened Heritage and Endangered Ecology in Varanasi

Pralay Kanungo

Abstract

Varanasi represents a unique confluence of river, religion, and city. An ancient city revered as a holy city by Hindus, Varanasi displays magnificent material heritage in the *ghats* (riverfront) built on the banks of the sacred river Ganges, where Hindu rituals like Ganga Aarti are performed every day. Varanasi has also been a symbol of Hindu-Muslim co-living for centuries, demonstrating the heritage of cultural and communal harmony in multiple arenas—from music to weaving. However, the core elements of this iconic river-city, which constitute the identity of Varanasi, are threatened today as inter-community relations become strained, material and cultural heritage are endangered, and the pure sacred river inches toward extinction due to severe pollution.

Keywords: Hinduism, heritage, Ganges, Muslims, harmony (*tana bana*), pollution

Introduction

Contemporary Varanasi, also known as Kashi in ancient times, and popularly called Banaras in medieval and modern times, is located in the sprawling river basin of the Ganges, the most sacred river of Hindus (Doron et al. 2015). Since ancient times this city has been prominent in terms of history, religion, culture, and commerce. In popular imagination as well as in textual depiction, Varanasi is a site of sacred pilgrimage (tirtha), a unique Hindu city of antiquity (Eck 1983). Varanasi is "quintessentially Hindu," "particularly ancient," as well as "universal, cosmic, and in a sense timeless and sacred

Padawangi, Rita, Paul Rabé and Adrian Perkasa, *River Cities in Asia*.
Taylor & Francis Group, 2022
DOI: 10.5117/9789463721851_CH06

space" (Gaenszle and Gengnagel 2008, 7). Without debating the city's claims of ancientness and Hinduness, it may be argued that contemporary Varanasi represents myths and memories, texts and practices, faiths and rituals, patronage and control, ideology and power relations, and political economy and governance, which took shape in pre-colonial/medieval period, was consolidated during colonial rule, and was carried forward in post-colonial India.

Though known as "the sacred city of the Hindus" (Sherring 2016) Varanasi is multi-religious and multicultural; the river-city had a long history of harmonious co-living among various communities. However, today, the tradition of co-living is threatened as intra-community relations among Hindus as well as inter-community relations between Hindus and Muslims, celebrated as *tana bana* (warp and weft) or Ganga-Jamuna *tehzeeb* (Ganga-Jamuna culture) in popular parlance (Raman 2010), become increasingly contentious. It is not just Varanasi's socio-cultural heritage, but also its age-old material culture that is being erased in the pretext of modernization, development, and beautification of the city. Moreover, the sacred river, worshiped for millennia for its pristine purity, has been contaminated with severe pollution and now confronts serious ecological degradation, which thereby threatens the survival of the river-city. In this context, the objectives of this chapter are: first, to understand how the interconnection between river, religion, and city constructs a unique identity and heritage of Varanasi; second, to discuss how religious traditions of the river-city become politicized and its material-cultural heritage trampled upon for the sake of development; third, to analyze the nature of community co-living in the city and explain the reasons for contestations and their impact on everyday life; and finally, to reflect on the deteriorating ecology that poses a serious threat to the survival of the river-city.

The chapter is divided into five sections. The first section briefly introduces the river, religion, and city—Ganges, Hinduism, and Varanasi—arguing how all three are integrally connected and define Varanasi's unique identity. The second section sharpens the discussion further by focusing on the riverscape, particularly spectacles like ghats (riverfront) and Ganga Aarti (rituals). The third section brings out how the interplay of religion, politics, and development threaten the age-old material-cultural heritage of Varanasi. The fourth section discusses the dynamics of intra-community and inter-community relations in the city, showing how social hierarchies, political economy, and communal mobilization adversely impact the shared heritage, making co-living contentious. And the final section raises a serious alarm on ecological degradation, which, cutting across communities, threatens the very survival of the sacred river, the city, and its citizens.

River, Religion, and City: "Cosmic" and "Iconic" Trinity

Rivers are prime locations of religious and cultural significance in South Asia (Feldaus 1995); the sacred river is supposed to be the meeting point of three worlds—heaven (*swarg*), earth (*martya*), and the underworld (*patal*). The Ganges, the most famous and sacred river in world history, the Nile perhaps being its possible rival, is the most revered river by Hindus and popularly called Ganga Maa (Mother Ganga). This river has given birth to many cities on its long and sprawling bank, of which Kasi/Banaras/Varanasi has always been distinct. In Hindu mythology, Ganga is a goddess, a celestial and earthly river (Ward 2010; Darian 1978; Doron et al. 2015), which came down to earth from the heavens. Since the force of its descent to earth would have been unbearable for the planet, Lord Shiva caught the river in his head to break her fall and eventually released a drop to flow as the River Ganga.

The myth of Ganga as a beloved consort of Shiva is found in inscriptions in Cambodia and in temples in Singapore (Narayan 2012, 427–428). The sacred water (Alter 2001) from the River Ganga is used in everyday Hindu rituals—from purification to death rites; bathing in the river is believed to purify all sins. Ganga transcends spatial boundaries as overseas Hindus can invoke the sacred Ganga in almost any waters anywhere; for instance, the Siem Reap River in Cambodia, the Ganga Talao lake in Mauritius, the River Aripo in Trinidad, and the confluence of three rivers in Pittsburgh.

As the twelfth-century text *Kashi Khand* mentions, the River Ganga, the patron deity Lord Shiva, and the sacred territory of Kasi together form the cosmic trinity. In the case of Varanasi, the river has been the anchor of Hinduism. Ganga's unique influence in the lives of Hindus around the world raises the importance of Varanasi, located in the middle Ganga Valley between the Varuna and Assi rivers, as a "supercharged" pilgrimage city (San Chirico 2012, 114–116). Known as Kashi ("luminous"), the City of Light, in ancient times, it is one of the oldest cities in the world, established around 3000 years ago; Kashi is a *tirtha*, a Sanskrit word meaning "ford." Significantly for Hindus, Varanasi is a crossing between the worlds of gods and humans. Kashi is Lord Shiva's chosen city; other city monikers attest to Shiva's bond: Kashi is Avimukta ("never forsaken").

Varanasi, the city's revived, post-Independence designation, is referred to in the Buddhist Jataka tales and in the Hindu epic Mahabharata. The city was identified in the Pali language as Banarasi, from which emerged the corrupt form of the name, "Banaras," by which the city is still widely known. Varanasi has long been a destination of pilgrims for the liberation of the soul. As the Sanskrit proverb proclaims, "Kashyam marnam mukti" (Death in Kashi is liberation). Even some non-Hindus wish their ashes to be immersed in Ganga after their death. For instance, former Australian cricketer Steve Waugh came to Varanasi to immerse the ashes of his friend Stephen in the holy Ganga to fulfill his dying wish (*The Financial Express* 2017).

For pilgrims, besides the river Ganges and Lord Shiva, the landscape of Varanasi is no less sacred. Thus, Ganga, Shiva, and Varanasi are intimately connected, forming an inseparable divine trinity. This trinity is not only "cosmic" but "iconic" as well. Religious icons are not essentially given, nor are they revealed, but they usually take shape as authorized socio-cultural constructs; once established, they foster religion in all its dimensions of experience, materiality, cognition, and action. Even artificial and natural objects can be referred to as religious icons if they trigger religious communication, including action and experience that is attributed with religious meaning (Knott et al. 2016). The iconic development of the trinity—Ganga, Shiva, Varanasi—confirms this understanding.

The Riverfront: Riverscape, Religioscape, and Cityscape

The riverfront at Varanasi, spread between the Varuna River in the north and the Assi River in the south, fuses the natural beauty of a living river with performative rituals and splendid architecture. Varanasi has a wonderful riverfront with some 84 ghats ("stone stairways"), constructed for religious bathing and rituals; an array of shrines, temples, and palaces rises tier on tier from the water's edge (Desai 2017). While walking along the ghats one can see pilgrims taking a "holy dip" in the river, making offerings to the goddess Ganga, performing rituals, and praying to the Sun god. Thus, the riverfront is a signifier of a Hindu space.

Representations of the riverfront as a timeless Hindu space have been found in indigenous discourses and practices. The *Kashi Khand* and Puranas described certain spots along the riverbank as *tirtha*, and Brahmins (priests) promoted a specific vision for this space through ritual practices like *Shanpuja* (ritual bathing and worship) and by consolidating the riverfront as a ritual and exclusively Hindu space. Indigenous representations treated the riverfront not only as a ritual space, but also as a spectacle of royal/elite patronage. The oldest recorded riverfront building, Bundiparkota ghat, was built by Rao Surjan in the sixteenth century, and the Maharajas of Jaipur constructed the Man Mandir and Zkanganwali Haveli on Panchganga ghat in the seventeenth century. Most of the buildings on the riverfront were private mansions.

After the installation of a Hindu king in Banaras in the eighteenth century, a number of Hindu royal and feudal elites from across the subcontinent, particularly the Marathas, invested in the material landscape of the riverfront. In 1735, Maratha Peshwa sponsored the construction of three ghats (Dashashwamedh, Manikarnika, and Panchganga); the project included the building of stepped stone terraces in order to provide easier access to the river. For ritual activities of the city Brahmins, Maha Brahma and Durga ghats were also constructed. Manikarnika ghat was a funeral ghat in the late eighteenth century and Scindia ghat was constructed in 1835. Some Muslim rulers patronized the construction of ghats as well. In 1735, Meer Rustam Ali, the governor of the province built the Meer Pushta on Meer ghat. Thus, the riverfront as a whole comprised a series of stone embankments that contained elite residences, monasteries, and temples. The patrons constituted a homogeneous group of elites: rulers/officials of princely states, landowners, and *mahants* (abbots) of powerful Hindu monasteries. Their objective was not only to display visibility but also to demonstrate authority vis-à-vis declining Muslim rule and in the face of rising colonial power.

Colonial discourse, following its own rationale, also contributed to the representation of the riverfront as a Hindu space as the colonial government was keen to sustain the riverfront's picturesque qualities as well as to retain it as an important space for public rituals. Colonial attention focused on the Dashaswamedah and Manikarnika ghats, the two principal "public" ritual ghats along the riverbank, as they were a great attraction for pilgrims and tourists. However, in the midst of this "Hindu" space stood the Dharahara Mosque, which was constructed by the Mughal ruler Aurangzeb by demolishing the Bindu Madhav Temple; the mosque remains a reminder, a signifier of Islamic iconoclasm in Varanasi.

Thus, the built environment of Varanasi, like the ghats, though presented as archaic, was constructed between the late eighteenth and early twentieth centuries. With its monumental architecture the built riverfront created a fusion of the riverscape and the religioscape, thereby constructing an image for the city as "Hindu." The patrons also periodically organized public spectacles, such as riverside festivals like Burwa Mangal, which not only helped the elites to garner social and cultural capital but also connected river, religion, city, and communities. The riverfront also gradually became a theatrical backdrop for rituals and spectacles.

Rituals and Performativity on the Riverfront: Ganga Aarti

For Hindus, *darshan* (literally, "seeing") is a key visual practice and an everyday ritual. In the spiritual sense, *darshan* means "seeing the sacred" or connecting the viewer with the divine. Hindus commonly use this term when they go to the temple to receive the "darshan" of a particular deity, that is to visually meet and connect with that god. Generally, the term signifies "auspicious sight" (Eck 1998, 104), and it applies to sacred images as well as to places or even persons deemed holy. The ghats, where the riverscape and the religioscape merge, have naturally become the ideal space to enact different visual practices. Among so many ghats, some have special significance for their ritual prominence; Dashaswamedah (called "the main ghat") is the most popular. Here the utmost celebration of vision takes place: Ganga Aarti, where the *darshan* of the holy Ganga is celebrated and brought to the stage every evening throughout the year. A few other ghats, like Assi ghat, also perform the Aarti.

Ganga Aarti (fire rituals), derived from Vedic fire rituals, is a religious ritual performed every evening on the ghats to worship the Ganges—it attracts pilgrims, tourists, and local residents. Undoubtedly, this ritual, which merges the sacred and the profane, has become a great public spectacle in recent years. The ceremony is conducted on a stage/platform and the steps of the ghat function as a sitting gallery for the devotees and spectators. This creates a perfect amphitheater overlooking the Ganges so that people can obtain unobstructed *darshan* and pay reverence to the river directly during the *aarti*. The river itself also acts as a privileged location to view this ceremony, as evidenced by the boats full of tourists and pilgrims parked on the waterfront.

The very programming, design, and performativity of this ceremony—from beginning to end—is impressive. The entire event—from the blowing of conch shells, ringing of bells, and singing of devotional songs, to the choice of instruments of liturgy, like cobra-shaped lamps, to the dressing up of the trained young priests in colorful, elegant regalia, to the style of conducting the rituals with the rhythmic and circular waving of hands holding vessels of burning incense and blazing fire lamps—has been choreographed with precision (Zara 2015). This performance combines elements of religious ritual, aesthetic intention, and entertainment and revolves around the sacred act of *darshan*, the divine vision of Ganga-ji. This spectacle is performed on a ghat which is an open public space. In this spectacle, the devotees and spectators are not merely passive observers: they are active participants praying, singing, clapping, and floating *diyas* (earthen lamps).

Besides regular Ganga Aartis, Varanasi hosts numerous religious festivals that add to the spectacles of the ghats. For instance, during the Ganga festival (dedicated to the Ganges), thousands of lamps are placed on the ghats and set afloat on the river. Brahmin families who traditionally controlled different ghats register their organizations (for example, Gangotri Seva Samiti), manage the ceremony, and collect donations from people. Some special *pujas* (worship) are also conducted for rich clients, overseas Hindus, and foreigners, raising large donations.

Transforming Heritage: Religion, Development, and Politics

The spectacle and performativity, tradition, and heritage have taken a distinct political turn in recent years. In 2014, Narendra Modi chose to contest the Varanasi parliamentary constituency. It was a strategic decision taken by the party of the Hindu Right, the Bharatiya Janata Party (BJP), that aimed to impact Hindu voters not only in Varanasi, the bastion of religious Hindus, but also across the largest and most populous state of the country, Uttar Pradesh (UP), and beyond. His supporters chanted "Har, Har Modi!", a modified version of a religious chant that venerates Lord Shiva, the reigning Hindu god of Varanasi. Launching his political campaign, Modi proclaimed: "Mother Ganga has summoned me" (Pagnamenta 2014). Modi performed Ganga Aarti in spectacular fashion and frequently invoked the Hindu religious symbols of Varanasi; media disseminated across the country and continents a live telecast of this spectacular ceremony and projected Modi as a devout Hindu worshipper of Mother Ganga and Lord Shiva.

Modi won with a huge margin of votes from the constituency and became Prime Minister. Later he launched the Swachh Bharat (Clean India) Mission in which Ganga and Varanasi became the focus of the campaign. No doubt, the Clean India campaign brought some infrastructural improvements to the city and the ghats and contributed to their beautification, thereby satisfying a section of his Hindu constituency; however, the real threat to the river, pollution, still looms large. The participation of world leaders in Hindu rituals like Ganga Aarti further added to the attraction of the spectacle, the aura of the prime minister and Hindu nationalism. With all this, Varanasi also became politically supercharged.

As Varanasi's representative in Parliament, Prime Minister Modi has undertaken developmental projects that aim to transform the city landscape. One of his pet projects, the Kashi Vishwanath corridor, has been launched to connect the riverfront to the shrine of Lord Shiva, the city's most revered religious site. Thus, from Lalita ghat to the Viswanath Temple, Varanasi has been clearing the way for a grand temple corridor by razing hundreds of houses, dismantling its oldest neighborhood, in order to improve accessibility for pilgrims by providing a direct pathway from the Ganges to the temple.

The objective of this project is to construct Vishwanath Dham, a large open complex around the Kashi Vishwanath Temple, by clearing around 45,000 square meters of space around the temple and creating dedicated pathways that will be 15 meters wide. The pathways, the project claims, will ease congestion at all four entrances to the temple (Agarwal 2019). Thus, instead of the narrow, crowded lanes, the project proposes to set up a hospital, a heritage library, resthouses, shops, cafeterias, helpdesks, offices for the temple workers, and a temple square. After all, the heritage city also has to be a "Smart City"; Varanasi has been designated as one of 100 Smart Cities that Modi plans to build across India.

The proposed construction will obviously redefine Varanasi's heritage, which is defined by its *galis* (narrow lanes); critics observe that construction of the corridor will likely destroy the very identity of the city. Moreover, the demolition has unearthed several ancient temples, statues, and historic buildings, raising concerns on how best to preserve this heritage while implementing the project. Although some people have lost their houses, shops, and livelihoods, protest seems to be muted in the city. While reasonable compensation has satisfied some affected residents, others, being Hindus and supporters of the ruling dispensation, accept it grudgingly. However, some voices of resistance still resonate. For instance, Swami Avimukteshwaranand, the head of the Sri Vidya Math, observes: "We are not against the corridor. But they have destroyed 15 or 20 temples and numerous idols. They had been there since ancient times. It is an attack on our faith" (Agarwal 2019). However, the administration denies that any temples have been destroyed and claims that due to the demolition of buildings, temples that remained hidden can now be visible: "All the temples that have come out will be protected. There will be more grandeur. They will be preserved and they will be a part of the corridor" (Agarwal 2019).

The project has deep political undercurrents as well. On 8 March 2019, while laying the foundation stone of the corridor, Modi claimed the project had freed Shiva from the clutches of the surrounding buildings. He described the event as a "liberation day" for the deity and added that for centuries, "enemies had their sight on this site ... many a time, it has been attacked" (Kumar 2019). This has obvious reference to the Mughal emperor Aurangzeb, who apparently demolished the Vishwanath Temple in the seventeenth century and constructed the Gyanvapi Mosque. At present, the mosque shares a boundary wall with the temple. The Hindu Right, which for some time has been staking claim to the mosque site, became emboldened by the prime minister's speech to revive their demands. Moreover, as some Muslim residents fear, the construction of this corridor will facilitate the entry of a large number of Shiva devotees from outside the area, enabling the Hindu Right to undertake a Babri-Masjid-like demolition in the future.

Communities and Contestations: Brahminical and Subaltern Hinduism

Hindu Varanasi is diverse and multi-layered. Social hierarchies and caste discrimination still afflict the Hindu social order. Brahmins (priests) constitute around 30 percent of Varanasi's population and play a dominant role in the ritual economy and institutions. Yet different Hindu *sanskaras* (ritual practices) require different agencies and actors besides the Brahmins. Thus, despite the existence of socio-economic hierarchies, interdependence between Brahmins and subalterns, by and large, guaranteed harmonious co-living among Hindus. As the post-Independence Indian Constitution abolished untouchability and discrimination, and guaranteed equal legal, social, and political rights to every citizen irrespective of caste, creed, or religion, marginalized Hindu communities started demanding recognition and rights. While struggling for social recognition and citizenry rights, these subaltern Hindus, at times, came into conflict with Brahminical Hinduism and the state. In this context, this section briefly discusses two marginalized Hindu communities—the *Doms* (Dalits), who belong to the lowest Hindu social order, and the *Mallahs* (boatmen)—and a stigmatized Hindu community, the *Aghoris*.

Varanasi celebrates death no less than life; it is believed to be a *Mahasmashan* (Great Cremation place). Hindus believe that one attains liberation from the cycle of reincarnations simply by dying here. Hence, in Orientalist representations Varanasi was not just a mystical, timeless, and picturesque place, but also odd, disquieting, and deadly. Cremation rituals are performed in the open-air along the ghats. *Doms*, formerly the "untouchable" castes, who occupy the bottom rung of the Hindu caste system, perform these rituals primarily to earn their livelihood through the "business of death" (Parry 1994).

According to Hindu mythology, cursed by Lord Shiva for theft, the *Doms* were assigned to become the keepers of the cremation flame. Today, about 35 *Dom* families live around Varanasi's two main "burning ghats"—Manikarnika and Harischandra—and earn their livelihood by charging a fee of 300 rupees ($5) per body for giving fire or "*agni*" to the dead (Shankar 2017). Earlier they used to provide the firewood and cremation materials. As nowadays people prefer to buy these materials from shops located outside the cremation grounds, they have lost access to the little extra money they used to earn from this sideline. After the corpse has been completely burned, children from the *Dom* families search through the ashes to find any ornaments left in the pyre. Even unused wood is collected and used in *Dom* households for cooking. The city government made attempts to regulate a standard rate for their work, but could not succeed. In most cases, *Doms* charge what the clients can afford. Although *Doms* are stereotyped as opportunists, ironically, this marginalized community are considered Hindu outcasts that are left with little choice but to take up this filthy profession. This so-called "business" prompts many to turn to alcohol and keeps their children deprived of education and respectable employment.

Mallahs (*Boatmen*)

Boatmen have long been an integral part of the river economy. With the development of the riverfront a new economic system evolved over time to meet the needs and fortunes of the boatmen. They have created rules and regulations, devised and regulated by the community, for ferrying pilgrims and tourists along the riverfront. Thus, the riverfront economy was characterized by a certain uniformity, mutuality, and moral conduct (Doron 2015). However, both upper castes and the state have often raised obstructions against this morally bound river economy by introducing numerous regulatory policies. For instance, the Ganga Action Plan (GAP), which was introduced by the state to curb pollution, proved to be detrimental to the boatmen's livelihood and economy as it banned fishing and put restrictions on the cultivation of land and sand-mining. The government's rationale was that fish helped in the decomposition of corpses and other organic material and thereby reduced pollution. The boatmen, however, argued that the real cause of pollution is the continual flow of factory effluent and urban sewage into the river.

The boatmen found in these policies a nexus between the state and the Brahmins: "The government has banned fishing in the name of cleaning the Ganga. There is a ban on fishing in the river on the pretext of cleaning the Ganga (*Ganga pradushan ke nam par*). Religious leaders and local pandas want to ban fishing because they consider it as killing animals (*jīv hatya*) in a holy place." While resisting this policy, they invoked their natural rights, arguing that it was their birthright (*hamara janm sid adhikar hei*) to fish and that they would continue to do so as their ancestors did. The community got organized and commenced open political action; boatmen associations wrote petitions, issued press releases, and organized protest rallies against these arbitrary government policies (Doran 2015, 325–340).

Interestingly, the boatmen fought for their rights by invoking their intimate relations with the sacred Ganges. They called themselves *Gangaputra* (Sons of Ganga)—a term usually used to refer to Brahmin priests—and employed devotional idioms to express their resentment towards the state policies. Referring to the river as their "mother," the boatmen argued that by virtue of their intimate relationship with the river "since time immemorial," they were entitled to certain rights and privileges like fishing, and the state had no right to deprive them of their natural rights. By claiming the title "*Gangaputra*," the boatmen contested the local, Brahminical forms of domination, and created a parallel national narrative imbued with symbolic and religious significance (Doran 2008).

Aghoris: *Living on the "The Wrong Side of the River"*

The *Kasi Khand* guarantees that those who die immediately across the river from Banaras will be reborn as donkeys, just as it promises spiritual liberation from all who die within the sacred interior of the city. So while pilgrims and priests scramble for a few square feet along the ghats, not a single Sadhu (Hindu religious mendicant) is willing to take up residence along the wide open spaces on the far shore of Mother Ganga. None, that is, except *Aghoris* (Barrett 2008, 101).

The *Aghoris* have been known as the most radical ascetics in India, who live naked on the cremation grounds, meditate on corpses, engage in cannibalism and coprophagy, and consume intoxicants out of human skulls. In recent years, however, they have shifted their practices from the embrace of ritually polluted substances to the healing of stigmatized diseases. Credit for this goes to Avadhut Bhagwan Ram who established an Ashram and a clinic on the "wrong side of the river" to treat people with leprosy and other skin diseases using modified Ayurvedic medicines. Built in 1961, Kusht Seva Ashram (KSA) at Parao has made an entry into the Guinness Book of World Records for claiming to have treated over two hundred thousand people with leprosy. With such intervention the *Aghoris* are no longer a socially marginalized sect. They have become a large, socially mainstream, and politically powerful organization. In the last few decades they have set up more than 150 ashrams and service centers, embracing thousands of middle-class people as well as millionaires and politicians. Apparently, leprosy did much more to legitimize the *Aghoris* than the other way round.

Wayside Hinduism: Another Site of Resistance

Varanasi has innumerable wayside Hindu religious shrines, old and new, venerating great and small gods of the Hindu pantheon. These icons, just as they are throughout India, are wide ranging and even trees and stones laced with vermillion are worshiped. These minor and marginalized shrines are not "sites of fatality," rather they represent popular/folk Hinduism promising protection and prosperity to the devotees. Usually, minor agencies and actors, individuals and small networks, organize and maintain the wayside shrines to raise some marginal resources. Sometimes, these become organized businesses, involving influential players, that grab prime urban land and convert minor shrines into major ones. As land becomes scare and expensive due to rapid urbanization, the state as well as the powerful try to evict these "illegal" encroachers of the public land and clear the wayside of religious structures. Sometimes, this becomes a hard struggle for survival for the people whose livelihood depends on the income from these gods.

One such site of resistance is Jogi Bir Baba's shrine at Rajghat. Birs (Brave) are brave folk heroes or martyrs who died a violent death. Varanasi is home to hundreds of shrines dedicated to Bir Babas (Coccari 1989). These Babas are ascetic guardian deities and are worshiped for protection. Folklore relates that Jogi Bir Baba would walk boldly in the forests of Rajghat, during dark nights when people would be afraid to go out of their homes. The priest—a former wrestler—at the Jogi Bir Baba shrine at the Rajghat site has been resisting eviction attempts made by a nearby wealthy spiritual institution. Thus, wayside Hinduism not only asserts its autonomy vis-à-vis Brahminical Hinduism, it very much resists attempts by the state and powerful economic interests to dislodge it from the city space.

Subaltern Hinduism: Resistance or Mainstreaming?

In the heartland of Brahminical Hinduism, non-Brahmins and the lower castes have acquired their own gurus, temples, and institutions. Ravidas, also called Raidas, a sixteenth-century mystic and poet who was one of the most renowned saints of the North Indian Bhakti movement, was born in Varanasi into an untouchable leather-working caste. His poems and songs often revolved around his low social position (Hawley and Narayanan 2006). However, his charisma compelled Brahmins to bow before him, and his poems were included in the Adi Granth, the sacred scripture of Sikhism. Consequently, a new subaltern religious movement was formed around him and a temple was built in Varanasi. His egalitarian teachings made him a figure of veneration and pride among the Dalits (the former untouchables); he was a Dalit icon who rebelled against the oppressive Hindu caste structure. Hinduism, however, has made efforts to co-opt him as another Hindu saint.

Political Hinduism continually makes conscious efforts to co-opt the marginalized as doing so has symbolic significance. Although the *Doms* of Varanasi are numerically insignificant, they are wooed by the Hindu Right to project an image of larger Hindu identity across the community. For instance, Varanasi's Dom Raja (King), the head of the *Dom* community, became one of those who proposed Prime Minister Modi's nomination for the 2019 parliamentary election. A pride for the community indeed!

Similarly, the *Mallahs*, though they fight against Brahminical domination in their everyday struggle, take pride in calling themselves *Nishads* who were connected to Lord Ram. They continuously invoke the role of Guharaj Nishad (King Guha, also known as Nishadraj) in the *Ramayana* and worship their ancestral king, who served Lord Ram by ferrying him and washing his feet. In 1979, a temple dedicated to Guharaj Nishad was erected on the riverbank and is now a center for community events. The encounter between Guharaj Nishad and Lord Ram is depicted across Banaras in wall paintings and pamphlets and the Hindu Right promises to build a 25-meter-tall grand statue of the Nishad King. This again evidences successful mainstreaming. More importantly, the Hindu Right government is sympathetic to the community's demand for Scheduled Castes status, which would ensure reserved positions in employment and education for the *Mallahs*.

The *Aghoris*, through their service, overcame the old stigma and by virtue of numerous reform measures became another credible Hindu sect with many devotees, elite patronage, and institutions. Kinaram, an *Aghori* saint, started the mainstreaming long ago by setting up Krim Kund (Pond), moving to the "right side of the river," and attracting a wide range of believers. Similarly, as resisting the state and the powerful becomes increasingly challenging, the small agents and actors of wayside Hinduism prefer to seek other, varied forms of patronage for their survival, and thus also become co-opted.

Muslim "Other" in "Hindu" Varanasi

Muslims have been living in Banaras since the twelfth to thirteenth century, and the city became an important Islamic center in the medieval period. Many old mosques in Banaras, like the Adhai Kangura, Ganj Shahida, and Abdul Razak Shah, and the tombs of Lal Khan and Gazi Mian bear testimony to the influential Muslim presence and their share in the city's culture (Kumar 1988). As the Indo-Persian culture of Awadh became a significant cultural characteristic of the region, Varanasi represented a rich tradition of Hindu-Muslim syncretism. For example, Meer Rustam Ali, the governor of Awadh, sponsored riverside festivals and Hindu festivals like Holi. Despite such syncretic traditions, spatial segregation of Hindus and Muslims existed on the riverfront, with ghats near the west end reserved exclusively for Muslims.

Like Hindus, the Muslims of Banaras are plural and diverse: Pathans and Ansaris, Sunis and Shias, Barelwis and Deobandis, and more can all be found in the city. Class division, and rich and poor are also in evidence. But a majority of the city's Muslims, the weavers, work in the famous handloom/silk industry, which existed long before the arrival of Islam in the eleventh century. Many of these Muslim weavers, called "julahas," are Hindu converts from the same caste (Kumar 1988, 49). Furthermore, Hindu money lenders, who occupied a central position in the weaving economy during the medieval period, were the only source from which poor Muslim weavers could borrow money to meet their financial and social needs (Raman 2010). As a consequence, this economic system created an interdependence among Hindus and Muslims and strengthened the bonds between them.

However, colonialism transformed the merchant-weaver relationship by creating a new hierarchy of mediators (Frietag 1989). Furthermore, de-industrialization restricted the sale of their product to a limited indigenous market. As *julahas* took part in the 1857 revolt against colonialism, they were soon labeled "bigoted *julahas*" and "fundamentalists" in colonial writings and classified as a particularly backward caste, in terms of being poor, illiterate, stupid, and at the bottom of the "pure-polluted" hierarchy (Pandey 1990, 19–28). Thus, for *julahas*, in the early twentieth century, "Islamization" became the means to promote self-dignity and for Muslims to attempt to gain a higher status. In the 1931 census they called themselves "Momin Ansari," tracing their descent as Prophet Mohammad's helpers (*nasir* in Arabic, plural *Ansari*); thus this title became a symbol of respect and dignity. Despite such assertions of identity, the Muslim weavers remained economically dependent on Hindu money lenders due to their lack of access to knowledge and capital.

In the post-Independence period, changes occurred in production relations which affected the interdependence between Muslim weavers and Hindu traders. While lower caste Hindus entered into weaving, Muslim traders graduated to trading. Furthermore, demand for mass production saw a shift from the handloom to the power loom and from hand designing to computer designing, greatly impacting the industry. The number of handloom weavers "dwindled to 20% of their original numbers, which by an informal count were a robust 50,000 or so" (Kumar 2016, 59). Certain government policies affecting the supply of yarn made the economics of producing Banarasi saris untenable and endangered the livelihood of the weavers. These new conditions strained economic and social relations between Hindus and Muslims and greatly reduced their age-old interdependence, thus facilitating Hindu-Muslim polarization and communal mobilization.

Varanasi witnessed its first "communal riot" between Hindus and Muslims in 1809. This violence apparently started at the site of Lat Bhairo and spread to the Gyanvapi Mosque area, later engulfing the city (Visuvalingam and Charlier-Visuvalingam 2006; Pinch 2012). Even after two centuries, it remains a contested site and uneasy calm descends when Hindus and Muslims assemble at the spot to offer prayers. The militant campaign by the Hindu Right for a Ram temple in Ayodhya unleashed communal riots also in Varanasi in the early 1990s. Instead of speaking of the old *tana bana* (Hindu-Muslim syncretism), popular discourse shifted to LoC (Line of Control) and "Mini Pakistan." Varanasi then became a target for "Islamist terrorists." On 7 March 2006, bombs were detonated at the Sankat Mochan Temple and the Varanasi Cantonment Railway Station, killing more than 20 and injuring more than 100 people. Modi's Hindu Right government has further fractured Hindu-Muslim relations in Varanasi.

Dying River, Dying City?

It is a great irony that while the Ganga, which has been worshiped by Hindus for millennia as a sacred river for its pristine purity and its sacred water (*Ganga Jal*), is used by millions of living and dying Hindus, it has also been allowed to degenerate into one of the most heavily polluted rivers in the world. The same devotees who worship the Ganges for its divinity and purity feel little or no guilt in abusing the river and making it impure. Every day about 60,000 people visit the ghats of the Ganga for their holy dip. For many of Varanasi's people, Ganga is considered their lifeline. But a sizable population that lives in slums along the riverbank and is dependent on the river for their livelihood and use it both for religious as well as routine needs, add further to the pollution.

Most of these people, who do not care about their actions, perhaps do not believe that the Ganga's water can be polluted; there seems to be a symbolic dissonance between religious and environmental worldviews (Alley 2015, 272–301). Interestingly, the pilgrim priests (*pandas*) of Varanasi make a distinction between purity and cleanliness, arguing that Mother Ganga can, unfortunately, become unclean (*asvaccha* or *ganda*), but that she could never be impure (*asuddha* or *apavitra*): "For these pandas, the river Ganga is a Goddess who possesses the power to absorb and absolve worldly impurities" (Alley 2015, 274)

A substantial source of pollution, however, comes from the sewage from cities, towns, and villages located along the river's banks, including Varanasi. This has had a deteriorating effect not only on the purity of Ganga's water, but also on the river system's entire ecology (Dayal 2016; Kothari and Bajpai 2017; Alley 1994). During British rule, the first sewage system was constructed in Varanasi to cater to a maximum population of 200,000. Though the sewage flowed into the river then, the locations for outlet pipes were carefully selected so as to not to pollute the water quality near the ghats in the city. By 2011 the city's population had reached 1.2 million. Due to this rapid population growth, the construction of settlements on the low-level lands along the rivers Assi and Varuna has had severe detrimental effects on those rivers. The Assi River is now nothing more than a drain through which only the sewage from the city flows and enters into the River Ganga.

Like many other rivers, the Ganges sustains diverse flora and fauna, which has helped the river maintain the purity of its water. However, the indiscriminate use of pesticides in agriculture has had a hazardous impact on fish as well as on other vertebrates. Furthermore, the indiscriminate killing of turtles by fishermen, especially soft-shelled turtles, has reduced the scavenging capacity of the river system. And direct and accidental trapping and habitat destruction due to various developmental activities, besides pollution, has pushed the Ganga River Dolphin to the verge of extinction (Dayal 2016, 57).

Various Indian governments—from Indira Gandhi's to Rajiv Gandhi's to Manmohan Singh's—have launched plans for cleansing the Ganges, and cleaning the city, but have not achieved success. In 1986, the Government of India passed the Ganga Action Plan (GAP), in which the Ganga was declared a national river, and in 2009 the National Ganga River Basin Authority was formed and a sum of 7000 crore rupees was approved for the Clean Ganga project (Dayal 2016). Prime Minister Modi, after coming to power, promised to make Ganga "*nirmal*" (clean) and "*aviral*" (free-flowing) and in 2015 his government hiked the budget to a mammoth 20,000 crore rupees for the rejuvenation of the Ganga over the next five years. This flagship "Namami Gange" project aims at the comprehensive cleaning and protection of the river. The government also planned to rope in the support of locals living on the banks of the river as well as grassroots organizations and urban local bodies. However, by early 2018, the government had spent only 20 percent of the budget (Koshy 2018). Even today, there is evidence along the riverfront of sewage flowing continuously from many drains as well as effluents from small-scale industries. Moreover, a lot of solid waste is either thrown into the river or into the sewage lines and/or open drains; or it gets dumped in the river via around 30 such point sources located along the ghats in the city. The continuously worsening quality of the Ganga's water shows that the various government plans to cleanse the river have failed.

Interestingly, in the context of the alarming pollution of the Ganges, the Indian judiciary, recognizing Hindu faith in the river, has granted Ganga the legal status of a "juristic person." The High Court of Uttarakhand consecrated the two rivers Ganga and Yamuna along the lines of Hindu idols, companies, and the Church, declaring: "the Rivers Ganga and Yamuna, all their tributaries, streams, every natural water flowing with flow continuously and intermittently of these rivers, as juristic/legal persons/living entities having the status of legal persons with all corresponding rights, duties and liabilities of a living person in order to preserve and conserve the rivers." Invoking the doctrine of *parens patriae*, the High Court appointed the director of the Namami Gange Programme and the chief secretary and advocate general of the state of Uttarakhand as persons *in loco parentis*, as the "human face to protect, conserve and preserve (the) Rivers Ganga and Yamuna and their tributaries." These officers have thus been entrusted with the onerous task of promoting the health and well-being of both rivers (Quasmi 2007, 25–26). Despite judicial intervention, the action plan to save the Ganges moves at a tardy pace.

Conclusion

Varanasi has for long had a unique identity as a sacred river-city, being the confluence of river, religion, and city. During its evolution from ancient to modern times, the city never lost its unique character and the bond between river, communities, and city always remained strong. In Varanasi, religious and cultural practices and everyday life and livelihood were intimately intertwined, thus ensuring harmonious co-living between communities. As the river had been the lifeline as well as the object of reverence, communities learned how to relate and regulate economy and ecology. The city, despite experiencing, sometimes mindless, destruction of its material and cultural heritage during its long historical journey, successfully recuperated the loss by reconstructing and redesigning the city landscape with the introduction of splendid architectural styles, such as those found on the riverfront. However, contemporary Varanasi faces multiple challenges. First, inter-community and intra-community relations are under strain and often turn conflictual as new power relations emerge with changes in political economy and the assertion of right-wing politics. Second, the city also faces challenges to its rich material and cultural heritage. It is ironic that this ancient city still struggles to find a place on the list of UNESCO World Heritage sites, primarily due to state apathy. Moreover, instead of preserving the heritage, the state tampers with and even destroys material and cultural heritage under the pretext of development and beautification. Finally, the anchor of this river-city, the sacred Ganges, faces severe threat, as every day it inches towards death due to unprecedented contamination. If the life-giver's life is vulnerable, then the survival of the river-city is also in jeopardy.

Acknowledgment

A DFG-funded Fellowship at Humanities Centre for Advanced Studies/ Kolleg-Forschungsgruppe (KFG): "Religion and Urbanity: Reciprocal Formations" (FOR 2779), Max-Weber-Kolleg, Erfurt University, made this research possible.

References

Agarwal, Kabir. 2019. "In Modi's Varanasi, the Vishwanath Corridor Is Trampling Kashi's Soul." *The Wire*, 8 March 2019. Available at https://thewire.in/politics/kashi-vishwanath-corridor-up-bjp, accessed 19 March 2019.

Alley, Kelly D. 1994. "Ganga and Gandagi: Interpretations of Pollution and Waste in Benaras." *Ethnology*, 33(2): 127–145.

Alley, Kelly D. 2015. "Separate Domains: Hinduism, Politics, and Environmental Pollution." In *An Anthology of Writings on the Ganga: Goddess and River in History, Culture and Society*, edited by Assa Doron, Richard Barz, and Barbara Nelson, 272–301. New Delhi: Oxford University Press.

Alter, Stephen. 2001. *Sacred Waters: A Pilgrimage to the Many Sources of the Ganga*. Gurgaon: Penguin.

Barrett, Ronald L. 2008. *Aghor Medicine: Pollution, Death, and Healing in Northern India*. Berkeley, CA: University of California Press.

Coccari, Diane M. 1989. "Protection and Identity: Banaras's Bir Baba as Neighbourhood Guardian Deities." In *Culture and Power in Banaras: Community, Performance and Environment, 1800–1980*, edited by Sandria B. Freitag, 130–170. Berkeley, CA: University of California Press.

Darian, Steven G. 1978. *The Ganges in Myth and History*. Delhi: Motilal Banarasi Das.

Dayal, Raghu. 2016. "Dirty Flows the Ganga: Why Plans to Clean the River Have Come a Cropper." *Economic and Political Weekly*, 41(25): 55–64.

Desai, Madhuri. 2017. *Banaras Reconstructed: Architecture and Sacred Space in a Hindu Holy City*. New Delhi: Orient Blackswan.

Doron, Assa. 2008. *Caste, Occupation and Politics on the Ganges: Passages of Resistance*. Surrey: Ashgate.

Doron, Assa. 2015. "Hello Boat! The River Economy in Banaras." In *An Anthology of Writings on the Ganga: Goddess and River in History, Culture and Society*, edited by Assa Doron, Richard Barz, and Barbara Nelson, 325–340. New Delhi: Oxford University Press.

Doron, Assa, Richard Barz, and Barbara Nelson (eds.). 2015. *An Anthology of Writings on the Ganga: Goddess and River in History, Culture and Society*. New Delhi: Oxford University Press.

Eck, Diana L. 1983. *Banaras: City of Light*. New York, NY: Columbia University Press.

Eck, Diana L. 1998. *Darsan: Seeing the Divine Image in India*. New York, NY: Columbia University Press.

Feldhaus, Anne. 1995. *Water and Womanhood*. New York, NY: Oxford University Press.

The Financial Express. 2017. "Former Australian Captain Steve Waugh Visited Ganga Ghat in Varanasi to Immerse the Mortal Remains of Friend Stephen." 8 March.

Freitag, Sandria B. (ed.). 1989. *Culture and Power in Banaras: Community, Performance and Environment, 1800–1980*. Berkeley, CA: University of California Press.

Gaenszle, Martin, and Jörg Gengnagel (eds.). 2008. *Visualizing Space in Banaras: Images, Maps and Practice of Representation*. New Delhi: Oxford University Press.

Hawley, John Stratton, and Vasudha Narayanan. 2006. *A Dalit Poet-Saint Ravidas*. Berkeley, CA: University of California Press.

Knott, Kim, Volkhard Krech, and Birgit Meyer. 2016. "Iconic Religion in Urban Space." *Material Religion*, 12(2): 123–136.

Koshy, Jacob. 2018. "Why Clean Ganga Project Has a Long Way to Go." *The Hindu*, 2 August.

Kothari, Ashish, and Shrishtee Bajpai. 2017. "We Are the River, the River Is Us." *Economic and Political Weekly*, 42(37): 103–109.

Kumar, Nita. 1988. *The Artisans of Banaras: Popular Culture and Identity, 1880–1986*. Princeton, NJ: Princeton University Press.

Kumar, Nita. 2016. "The Weavers of Banaras." *Economic and Political Weekly*, 51(53): 59–63.

Kumar, Sushil. 2019. "How Modi's Kashi Vishwanath Corridor Is Laying the Ground for Another Babri Incident." *Caravan*, 27 April. Available at https://caravanmagazine.in/religion/how-modi-kashi-vishwanath-corridor-is-laying, accessed 5 May 2019.

Narayan, Vasudha. 2012. "Ganga." In *Encyclopedia of Global Religion*, edited by Wade Clark Roof and Mark Juergensmeyer, 427–428. Thousand Oaks, CA: Sage.

Pagnamenta, Robin. 2014. "Mother Ganga Has Summoned Me, Modi Tells Crowds." *The Times*, 10 May. Available online at https://www.thetimes.co.uk/article/mother-ganga-has-summoned-me-modi-tells-crowds-sfff2zzpftt, accessed 17 January 2022.

Pandey, Gyanendra. 1990. "The Bigoted Julaha." *Economic and Political Weekly*, 18(5): 19–28.

Parry, Jonathan. 1994. *Death in Banaras*. Cambridge: Cambridge University Press.

Pinch, William. 2012. "Hiding in Plain Sight: Gosains on the Ghats, 1809." In *Banaras: Urban Forms and Cultural Histories*, edited by Michael S. Dodson, 77–109. Delhi: Routledge.

Quasmi, Bushra. 2007. "Rivers as Legal Persons: A Regressive Step." *Economic and Political Weekly*, 42(30): 25–26.

Raman, Vasanthi. 2010. *The Warp and the Weft: Community and Gender Identity among Banaras Weavers*. New Delhi: Oxford University Press.

San Chirico, Kerry. 2012. "Banaras." In *Encyclopedia of Global Religion*, edited by Wade Clark Roof and Mark Juergensmeyer, 114–116. Thousand Oaks, CA: Sage.

Shankar, Priyanka. 2017. "Doms of Varanasi Make a Living among the Dead." Available at https://www.reuters.com/.../india-doms-varanasi-cremation-idUS, accessed 10 October 2018.

Sherring, M.A. 2016. *Benares: The Sacred City of the Hindus* [*1868*]. Varanasi: Pilgrims Publishing.

Visuvalingam, Sunthar, and Elizabeth Charlier-Visuvalingam. 2006. "Bhairava in Banaras: Negotiating Sacred Space and Religious Identity." In *Visualizing Space in Banaras: Images, Maps and Practice of Representation*, edited by Martin Gaenszle and Jörg Gengnagel, 95–126. Wiesbaden: Harrasowitz Verlag.

Ward, William. 2010. *A View of the History, Literature, and Religion of the Hindoos: Including a Minute Description of Their Manners and Customs, and Translations from Their Principal Works*, 264–280. Cambridge: Cambridge University Press.

Zara, Christina. 2015. "Rethinking the Tourist Gaze through the Hindu Eyes: The Ganga Aarti Celebration in Varanasi, India." *Tourist Studies*, 15(1): 27–45.

Biography

Pralay Kanungo is Guest Researcher at Leiden University. He was DAAD Guest Professor at University of Heidelberg (2019-2020) and Fellow at Max-Weber-Kolleg, Erfurt (2018-2019) and ICCR Chair of Contemporary India Studies at Leiden University (2013–2018). He specializes in Hindu nationalism, religion and urbanity, and Indian politics. Kanungo authored *RSS's Tryst with Politics* (2002) and co-edited *The Algebra of Warfare-Welfare* (2019).

7 River-Edge Relationships

A Comparison of Riversides in Indonesian Cities

Michaela F. Prescott

Abstract
Although the poor condition of rivers in Indonesian cities is significant, there is limited documentation and comparison of river improvement projects specific to socio-spatial factors. Furthermore, there is a need for practical examples of projects to support policy-making, design, and implementation. This chapter describes a comparative analysis of qualitative aspects of river improvement in Indonesia, based upon fieldwork documenting material evidence of the reconfiguration of the edges of four urbanized rivers. This evidence was classified based on aspects of the physical environment influencing the river-edge relationship. The findings illustrated the potential of neighborhood-centered projects to have positive environmental impacts. The chapter concludes with a series of forward-looking recommendations for the design of river-city relationships.

Keywords: urbanized rivers, river improvement, socio-spatial factors, policy and design, river-edge relationships

Introduction

Rivers, along with other waterways, are key elements within a wider environmental water system and continue to provide a range of services to human and non-human inhabitants. The past decades of immense demographic and economic growth within Asia have seen the region's rivers heavily compromised (see also Chapter 1). Like the rest of the world's rivers, Asian rivers have collected and carried the waste of adjacent settlements (De Meulder and Shannon 2013; Vollmer and Grêt-Regamey 2013). As the materials used in daily life have changed (for example, from organic to inorganic) and increased exponentially, with the volume of waste and contaminants they have surpassed the capacity of rivers. Cities across Asia are reaching what *The*

Padawangi, Rita, Paul Rabé and Adrian Perkasa, *River Cities in Asia*.
Taylor & Francis Group, 2022
DOI: 10.5117/9789463721851_CH07

Economist has termed "environmental turning-points" (2013) as the impacts of these patterns are revealed through events which illustrate the extreme environmental degradation that has been sustained (De Meulder and Shannon 2013; Lucas and Djati 2007). Indonesia, like other developing nations, has been repeating a pattern established by other countries including Britain, the US, and Japan: "develop now, clean up later" (*The Economist* 2013; Nakamura et al. 2006). For example, environmental pollution was not perceived as a problem by East Javanese people until the mid-1970s and the Surabaya River, the main source of drinking water for the region, was a dumping ground for rubbish and untreated waste from factories along its banks (Lucas and Djati 2007, 322).

According to a survey undertaken in 1990, there are some 5,000 major rivers in Indonesia, and some 66,000 tributaries (Sukardi et al. 2013). Lakes and rivers in Indonesia, particularly within densely populated areas, are still heavily polluted, serving as dumping grounds for liquid industrial waste, solid waste, and sewage (Prihandrijanti and Firdayati 2011; Lucas 1998, 181). In Indonesia, those unable to afford to enter the formal housing market historically have occupied and settled in a range of available spaces, including coastal or marshy areas, unused lots, and spaces along railway tracks, canals, rivers, roads, and under bridges (Abeyasekere 1987). These low-income residents value the services that these vulnerable environments provide. In the context of riverine neighborhoods, households may rely on multiple environmental services (e.g., direct sanitary use, recreation, harvesting plants, groundwater use, solid waste disposal, sewage disposal). Meanwhile, residents reliant on these services are also at greater risk of water-related hazards, service disruption (e.g., a decline in water quality), and, potentially, eviction (Vollmer and Grêt-Regamey 2013, 1545). Vulnerable urban settlements are particularly susceptible to water-related hazards caused by extreme weather conditions (Sakijege et al. 2014) and compounded by development pressures. For example, in 2017 at least 640 floods hit Indonesia between January and October, at that time the most frequent on record with the largest loss of life and number of missing people (Antara News 2017). It is therefore significant that riverside transformations and activities within this geographical context are studied and reflected on, for the betterment of river-city relationships through design transformation.

Practice and Policy

Throughout Indonesia riverside settlement is not a new phenomenon. Despite the inclusion of guidelines for the preservation of the river edge—or borderline (*sempadan*)—in government regulations for some decades already, in practice these guidelines have not been consistently upheld (Kumorotomo et al. 1995). *Kampungs*, self-initiated urban settlements (Tunas and Peresthu 2010), have been addressed through urban upgrading programs such as the Kampung Improvement Program (KIP), even during the Dutch colonial period. However, upgrading programs were standardized across *kampungs*, and improvements implemented within riverside areas were not riverside habitat-specific and generally did not include waterfront transformations or account for local activities or environmental services. These programs consisted of primarily technical interventions, focused on public toilets, access (i.e., pavements), drainage, and solid waste management. This technical focus may have contributed to the lack of success in resolving key issues pertaining to deteriorated conditions such as environmental awareness and empowerment (Yossi and Sajor 2006).

The Prokasih (Program Kali Bersih or Clean Rivers Program) was initiated by the government of Indonesia in 1989 to improve the water quality of rivers. The program aimed to reduce the pollution load from industrial activities through a two-stage process over ten years. This focused on identification of polluters, control measures and treatment, and standards and regulations in the first five years; then shifting to policing and mitigation (Asian Development Bank 2007). There was reportedly a lack of compliance as a result of conflicting development interests and a failure of political will (Warren and Elston 1994, 8); most discharges were non-conforming, few sanctions or fines were imposed, and water quality continued to deteriorate (Asian Development Bank 2007). In some cases, the program implementation at a municipal level was even perceived by the public as ceremonial—with events coinciding with national holidays—or simply rhetoric (Raharjo 2010).

River policies, such as Government Regulation No. 38/2011 regarding Rivers—referred to locally as Peraturan Pemerintah 38/2011 tentang Sungai (Republik Indonesia 2011)—omit the characterization of the river border and promote functional river corridors. In the following decade (2011-2021), this was largely enacted through a piecemeal channelization (normalization or *normalisasi*) program and the removal of riverside settlements. This practice falls short of the current model of multifunctional river corridors (Greco and Larson 2014) that is practiced around the world, and provides constituents with a broader range of ecosystem services. In 2006, Yossi and Sajor identified the coherence of policies across national, provincial, and municipal levels as a primary challenge in both the management of rivers in past decades, and future development of rivers and their waterfronts in Indonesian cities (2006). The management of conditions along major rivers has historically been complicated by contradictory regulation, relating to private land ownership and state control, including between the National Land Agency's (Badan Pertanahan Nasional) legislation of private ownership rights and Government Regulation No. 38/2011 regarding Rivers, which states that rivers and borders are controlled by the state (Prescott and Ninsalam 2020, 75). Furthermore, those rivers crossing provincial administrative boundaries would be managed by the national government, not local authorities, and as a result, river management has historically not integrated across upstream and downstream reaches. Yossi and Sajor (2006) recommended the following critical actions: local-river-specific planning (rather than a national regulation that generalizes the characteristics of all rivers); and collaboration and coordination between government authorities and external stakeholders.

Strong examples of community-led action around environmental advocacy and education emerged as a result of the government's limited financial resources and lack of coordination across national, provincial, and municipal boundaries (Prescott and Ninsalam 2020, 70). It was only in 2019, when the House of Representatives passed a new law on water resources (Law no. 17, 2019), that key legislative shifts relating to integrated water resources began to occur (Prescott and Ninsalam 2020, 74). The new law prioritizes public access to Indonesia's fresh water and is reinforced by the government's "naturalization" (*naturalisasi*) program, which is focused on social and cultural change through environmental education, clean-up, and eco-tourism. Increasingly, the government has begun to partner with community-based organizations to action the management of domestic water resources, building on exemplary community-led projects. The new law is actioned through a series of president-directed River Task Forces (SatGas, whose membership includes ministries, academics, the army, and community groups) to ratify River Action Plans which identify priority river management actions for the next five years. Although recent legislative changes have improved collaboration and coordination between different government authorities and external stakeholders (donor foundations and research institutions) for urban river management, approaches to "local-river-specific planning" are still under development.

Landscape Services

To support local-river-specific planning, this chapter takes "landscape services," as defined by Termorshuizhen and Opdam, as a conceptual framework. Landscape services are defined as a "bridge between landscape ecology and sustainable development" (Termorshuizhen and Opdam 2009,

1041), and contextualize ecosystem services to the local landscape scale. Landscape may be directly "associated with people's local environment, with the place for which they feel responsible, and with distinct spatial elements that they can change to improve the ecological, social, and economic value" (Termorshuizhen and Opdam 2009, 1043). Landscape services are therefore used as a specification of ecosystem services in light of landscape development involving local actors such as residents and authorities. The chapter also builds on the premise in landscape development that people are part of the landscape and that landscapes are changed for their benefit (Antrop 2005), and focuses largely on the following landscape services (Hermann et al. 2014):

- regulation, related to the capacity of cultural landscapes to regulate essential ecological processes and life support systems;
- habitat for wild plants and animals;
- provision, relating to the supply of natural resources such as "edible wild plants and animals," "raw materials," "genetic" and "medicinal resources";
- information, meaning all services referring to spiritual enrichment, cognitive development, recreation and aesthetic experiences; and
- carrier services, which describe the capacity of landscapes to provide suitable substrate (soil) for activities such as "cultivation," "habitation," "tourism facilities," "energy conversion," and "transportation."

A mixed-methods approach (literature review, field observation, convivial interactions) is employed to identify transformations and activities within urban and suburban riparian corridors following Zeisel's framework for exploring "traces" (2006) and Prominski et al.'s investigation of river-edge typologies (2012). In so doing, a spatially explicit methodology that examines the physical reconfiguration of riverine landscapes in Indonesia as a result of human intervention and implemented policies is presented. These studies help to identify the landscape services and relationships important to residents' quality of life along Indonesian rivers. By linking physical conditions of riversides to less tangible social and cultural benefits, alongside more technical benefits relating to conveyance and mitigation, research can be tailored to issues important to end-users and a more holistic accounting of the potential benefits of different riverbank typologies is achieved.

This introduction has demonstrated the potential for integrating physical and sociocultural factors in the assessment of urban riverside improvement. In the next section a set of case studies of riverine landscape transformation in Java, Indonesia is established and a methodological and analytical framework for comparison and contrast is developed. The discussion then illustrates how observation and spatial analysis can be used to examine transformations to densely riverine landscapes specific to sociocultural and spatial factors. This is concluded by a discussion of the implications of spatial measures and recommendations for future spatial reconfiguration, and design of riverine landscapes in Indonesia.

Materials and Methods

This chapter employs primary data that was collected in 2013 through convivial interactions with riverside residents and urban practitioners, structured field observations, and a literature review. The primary data collection was directed by the following questions, derived from literature on environment-behavior studies:

> How do environments create opportunities for people? Where do people and their surroundings impinge on each other? Where do they limit each other? How do people use the environment as means to an end? And to what ends? What design skills do people have? How do they manipulate their surroundings? How do people change environments to meet their needs? What takes place in particular settings? (Zeisel 2006)

It is important to note that "design" is here defined as intentional landscape change (Zeisel 2006; Nassauer and Opdam 2008). The literature review included English-language publications, and, where available, Indonesian-language regulations and publications (translated into English). The chapter describes a practical application of a comparative analysis of qualitative factors in river improvement. The analytical framework is focused on the integration of spatial (physical) and sociocultural considerations (Prescott and Ninsalam 2016). Although the study is concerned with rivers in Java, the overall approach could be transferable within Indonesia or internationally. The methodological framework is based on data availability and the spatial scale of the local landscape. The case study selection was established using geographical region and landscape types as organizing characteristics, informed by the case study method for landscape architecture (Francis 2001).

The fieldwork and reviewed details considered the rivers' physical environment, including spatial relationships of the river edge, settlement and ecology, and the sociocultural aspects, including community and organization, recreation and access. Based on a grounded theory approach, data was recorded in the field using photographs, cross-sectional illustrations, and annotations. These were then translated into socio-spatial indicators—or traces (Zeisel 2006)—and linked to the related spatial types (Prominski et al. 2012) and the landscape services they provided. The socio-spatial indicators influencing the selection of spaces for discussion included: access, crossing, open space, river usage, and spatial structure (including orientation and built form).

Case Study Context

Empirical evidence is drawn from four urban riverine case studies in Java, Indonesia (Figure 7.1). The case studies selected include rivers in Central and West Java: the Code and Winongo (Yogyakarta); the Cikapundung (Bandung); and the Bengawan Solo (Surakarta). The rivers under study are primarily short rivers, less than 50 kilometers in length—excepting the Bengawan Solo—and pass through densely settled, urban and sub-urban areas. In the case of the Bengawan Solo, the portion studied is that which passes through the urban and sub-urban areas of Surakarta, a segment around 15 kilometers in length. Sites in both the urban and sub-urban reaches of the river were visited for each case, providing background to the challenges particular to each river. Each river passes through a number of districts and/or sub-districts.

The Kali Code and Kali Winongo are located in the special administrative region of Yogyakarta (Daerah Istemewa Yogyakarta). They are the downstream tributaries of the Opak River, and bisect the city from north to south. Located at the heart of the city, the Kali Code is considered the most important of the rivers crossing the province and has been described as "the center of life" for the communities living along it (Toumbourou 2009). It flows from the active Mount Merapi volcano in the north of the province to Parangtritis beach in the south, and during eruptions regulates the lava flows through the city (Seftyono 2012). Informal settlements along the Kali Code were affected during the eruption of Mount Merapi in 2010, which caused a cold lava flow into the neighborhoods along the banks of the river (Asian Development Bank 2012). It also has provision, information, and carrier services, functioning as a source of water, and a focus for cultural and religious activities such as the traditional "Merti Code" river-cleaning activity, as well as ecotourism (Seftyono 2011). While the city spreads in all directions from the Sultan's palace, the core of the modern city is to the north, centered on Dutch colonial-era buildings and the commercial district. This city, and surrounding area, had great significance through the Mataram Empire, Majapahit Kingdom, and Mataram Sultanate. Yogyakarta's image and identity is strongly shaped by Indonesian culture and tradition (Gill 1993). As the only Indonesian royal city still ruled by a monarchy, and with people of predominantly Javanese ethnicity, it is regarded as an important center for classical Javanese fine arts and culture.

The Cikapundung River in Bandung City is one of the upstream tributaries of the Citarum River. The total length of the river is 28 kilometers, of which 15.5 kilometers pass through the city itself. Although the local people are predominantly Sundanese, Bandung has been strongly influenced by its Dutch colonial history and is known for its blend of cultures and prototypes of Indo-European architecture and town planning (Gill 1993). While in the Dutch colonial period the Cikapundung River Valley was referred to as "one of the few examples of natural scenery having been converted into scenes of natural beauty" (Toelichting 1938 cited in Reerink 2011, 28), the Cikapundung now suffers a range of environmental problems (Farida et al. 2007). It has a regulatory function, servicing natural drainage and flushing of untreated wastewater from surrounding, primarily urban, areas. It also provides untreated drinking water and hydroelectricity, and carries a range of ecotourism opportunities. The government has initiated programs to increase the setback between the existing informal settlements and the river's edge in order to improve safety, access, and appearance (Windarti 2016).

The Bengawan Solo River is the longest river in Java Island, stretching to 600 kilometers and flowing from the Sewu Mountains (Thousand Mountains) in the southwest of Surakarta in Central Java Province to the Java Sea in the northern part of Surabaya through East Java Province (see also Chapters 2 and 4). The river used to serve as a means of transportation and commerce along Java's interior, with 44 ports lying between Wonogiri in the upper reaches and Gresik downstream (Adi 2012). The river indeed witnessed a long historical passage extending from the early Mataram, Majapahit, Demak, Pajang, and Islamic Mataram kingdoms to the Surakarta Sultanate (SMERU Research Institute 2011). The waterway was significant in Surakarta between the thirteenth and nineteenth centuries, when it was a major thoroughfare for agricultural produce and merchandise. Like Yogyakarta, its people are predominantly of Javanese ethnicity. The Bengawan Solo River basin, which is the longest and the largest river basin on Java Island, drains approximately 16,000 kilometers of the island (Takeuchi et al. 1995).

Figure 7.1 Map of comparative case study illustrates selected rivers in Java, in comparison to the location of the Ciliwung River in Central Jakarta

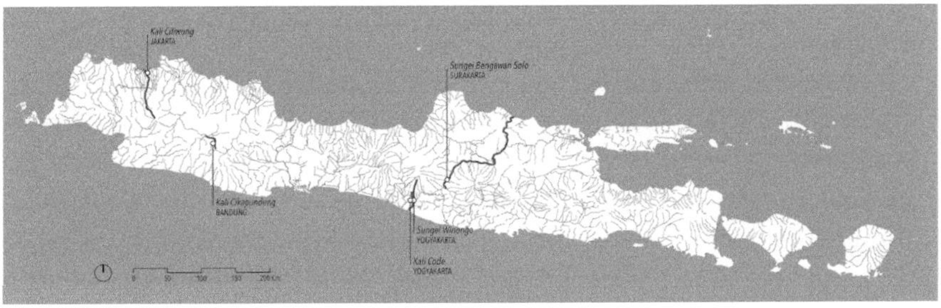

Among other functions, it provides water for irrigation and hydroelectricity, and is used for flood management. Its significant catchment size has meant that cities sited along its banks and within its floodplains have suffered high flood risk.

With this brief introduction to these four rivers and their cities it is clear that each river and its respective city are unique. Their particular urban development histories, the sociocultural context, environmental and cultural significance mean that local-river-specific planning that takes into account the particularities of each river and its situation is necessary. At the same time, comparative case studies are particularly important to questions of human-environment interaction since they combine both case study and comparative research traditions and thus extend the depth of the individual case study across several instances. The findings are limited by the number of case studies and instances—or moments along rivers—that could be observed within this exploratory field investigation, and are not, by any means, deep ethnographic accounts across seasons. This said, in taking this approach, those cases that have been encountered can be compared and contrasted in order to understand how the surrounding sociocultural and spatial context influences the sustainability of an intervention and how better to tailor that, or other, interventions to their specific contexts to achieve the intended outcomes.

Results and Discussion

The following discussion is focused on how sociocultural and spatial changes have affected the activities occurring along the waterfront, which are seen as indicators. Furthermore, the potential range of landscape transformations within a situation are affected by the physical environment: including types of adjacencies—such as settlements, agriculture, industry, and forestry or vegetation—and bank conditions. From looking across these case studies, a series of sociocultural and spatial measures have been identified. Sociocultural measures have included: (1) empowerment, facilitated through neighborhood involvement throughout the upgrading process; and (2) respect for environment, facilitated through activities, programs, and groups (Table 7.1). Spatial measures have included: (1) physical improvements, such as reorienting the neighborhoods toward the river; and (2) the introduction or improvement of amenities and facilities. The discussion is focused on how such measures have affected the activities occurring along the waterfront.

Based on the analysis of the interviews and structured observations, the results are structured in two parts: sociocultural and physical. First, the sociocultural aspects of measures that have been taken are discussed, followed by the physical, or spatial, measures that have been taken. Examples from each case study are compared and contrasted.

Sociocultural Aspects

Specific to riverine vernacular settlements, sociocultural aspects of landscape transformations included: the empowerment and involvement of communities; respect for environment (including stewardship, green infrastructure, and environmental campaigns); socio-economic activities; and recreation and access to the river. Table 7.1 details the types of activities or transformations that these entail, and the prevalence within neighborhoods included in the case study. While tenure regularization and development programs were components of some of these case study improvements, these will not be discussed in detail as they are not the focus of this review and have been discussed within the literature already. Socio-economic and domestic activities, while not being directly relatable to improvement strategies in these case studies, were observed to transform the riverine landscape in a variety of ways and are thus included within the analysis. While all activities are influenced by both sociocultural and spatial factors, recreation and access is largely defined by spatial factors and although it is included in this table it is described further on.

Table 7.1 Sociocultural and physical aspects of riverine landscape transformation

Variables			Case study				
			Cikapundung River, Bandung	Sungai Winongo, Yogyakarta	Kali Code (Kampung Code), Yogyakarta	Kali Code (Tegal Panggung), Yogyakarta	Bengawan Solo River, Surakarta
Spatial	Orientation (toward the river)		X	X	X	X	
	Access	Continuous walkway (along river)	X	X	X	X	X
		Facilitated at-grade	X	X	X	X	X
		Across (bridge)		X			
	Expansions	Foreshores (in river)	X	X	X	X	X
		Open spaces	X	X	X	X	X
	Walls and barriers	Embankment					X
		Half-wall	X	X	X	X	
		Wall (full-height)	X				
Social	Empowerment / involvement	Neighborhood-driven improvement	X	X	X	X	
		Involvement of community in design generation			X	X	
		Involvement of community in relocation process					X
		Involvement of community in maintenance	X	X	X		
	Respect for environment	Environmental stewardship	X	X	X	X	X
		Green infrastructure	X	X			
		Environmental campaigns	X	X			X
	Socio-economic and domestic activities	In river					X
		Along riverbank	X				
	Passive and active recreation opportunity	Facilitated access along river	X	X	X	X	X
		Facilitated access to river (at-grade)	X	X			
		Community facilities and spaces (adjacent to river)		X	X		X

Programs and activities relating to the river and riverside spaces

Empowerment and Involvement of Communities

It is well established in the literature that "people's active participation in designing, developing, and managing the physical settings of their communities can contribute to individual and collective empowerment" (Feldman and Westphal 2000, 109). Empowerment can be defined as a "process aimed at consolidating, maintaining, or changing the nature and distribution of power in a particular cultural context" (Morgan and Bookman 1988, 4). Positive examples of riverine neighborhood improvement programs in Indonesia are often focused upon community-based management. This reflects a development paradigm which is "people-centered, participatory, empowering, and sustainable; placing emphasis on local autonomy in the decision-making, self-reliance, direct participation, and social learning," and is characteristic of other positive examples in the region (Tahir et al. 2011, 839). Examples such as Kampung Code (and Stren Kali Surabaya, although this case study was not considered here) illustrate the value and success of neighborhood-driven improvement, sustained by the mutual support of residents. In the cases reviewed in this chapter, this involvement took several forms, including:

- neighborhood-driven improvement, whereby physical actions indicate readiness and mutual support from residents willing to improve the look of the village and take on a position of stewards of their environment (Bandung; Yogyakarta-Kampung Code, Sungai Winongo);
- involvement of communities in design generation, whereby communities propose spatial improvements based on their understanding of the local landscape (Yogyakarta-Kampung Code, Tegal Panggung);
- involvement of communities in the relocation process, whereby communities are involved in decisions about their place of residence and made aware of benefits of landscape changes (Surakarta);
- involvement of communities in maintenance (Bandung; Yogyakarta-Kampung Code, Tegal Panggung);
- and, involvement of communities in larger-scale physical restructuring of urban areas adjacent to rivers which can promote a sense of ownership or responsibility toward the ongoing maintenance and care of a site. Ultimately, this may support the ongoing sustainability of improvement projects. As in the case of Surakarta, making local situations visible through tools such as mapping can help improve the relevance of other assistance or development programs.

Respect for Environment: Stewardship, Green Infrastructure, and Environmental Campaigns

Environmental stewardship was revealed through physical signs of care and ownership (Nassauer 1995) throughout the case studies. While some neighborhoods had clearly defined collective environmental programs and campaigns, in others stewardship was evidenced mainly at the household level. This might be supported collectively and visible at the neighborhood scale. Consistently, the case studies in Yogyakarta, Surakarta, and Bandung revealed that neighborhoods had an interest in nature-based infrastructure— such as networks of green space that conserves ecosystem values and functions at the same time as providing benefits to human populations (Benedict and McMahon 2002)—and environmental campaigns more broadly. The involvement of NGOs and environmental groups was observed to be an important supporting factor. This can be linked to a broader contemporary movement within Indonesia to address gaps in servicing neighborhood wastewater treatment through decentralized and nature-based systems (Ramírez-Lovering et al. 2018; Prihandrijanti and Firdayati 2011).

In Bandung environmental groups implemented greening programs to improve neighborhoods directly along the river, as well as the river corridor itself. For example, in the reaches of the river passing through the city center, an NGO encourages neighborhoods to participate in greening campaigns, and brighten up narrow walkways and large walls with painted murals. Meanwhile in the upstream area another NGO has implemented a biofilter using *Vetiver sp.* to treat the domestic wastewater of a nearby village, and a third works with volunteers to revegetate the river edges, as well as educate. In Yogyakarta municipal schemes for river improvement are disseminated through education programs within riverside communities, in terms of individual behavior, such as garbage disposal (also described in Toumbourou 2009) and physical measures that can be taken within neighborhoods, such as *biopori* were observed. In this instance the *biopori*—a local method originating in rural areas of Bogor, where shallow bore holes are filled with organic matter to increase groundwater recharge and reduce storm run-off (Vollmer et al. 2015)—were simply holes drilled through hard surfaces such as drain covers to increase permeability. In Surakarta, following devastating floods in 2007, the relocation of 1500 families from an area adjacent to the river (Scheuer 2014) made way for a public park, urban forest, and floodplain. The local administration perceived the urban forest's ongoing maintenance as the responsibility of the city, and as an important step toward the city taking stewardship of and feeling a sense of ownership of the park. Since

Figures 7.2 and 7.3 Home gardens, either formal or informal, are physical signs of care and ownership of the riverside environment (Photos by the Author, 2012)

its completion, cleaning and maintenance events have been held involving the military, NGOs, and the public. A variety of cultural events have been arranged by the municipality to enhance the city's relationship with the river (such as the annual *getek* or raft parade).

Case study neighborhoods along the Code and Winongo rivers of Yogyakarta provide valuable examples of neighborhood-based improvements. Kampung Code illustrates the value of "bottom-up" and self-sustaining neighborhood improvement, whereby the community continue to maintain their environment and adapt to environmental changes. Household greening in the form of small-scale home-gardens (*pekarangan*) was observed in both formal and informal expressions in Bandung and Yogyakarta (Figures 7.2 and 7.3). *Pekarangan* are a traditional home-garden system in Indonesia that establishes and promotes greening and species diversity. Research has shown that *pekarangan* can contribute to a range of ecosystem services, such as carbon sequestration, water resource management, and urban biodiversity, along with reducing household food expenses (Arifin et al. 2014; Arifin and Nakagoshi 2011), and as such are a relatively achievable demonstration of respect for the environment at the household level with multiple co-benefits. In some neighborhoods, where there was less space available for gardens, space was found for potted plants along the riverside. In Kampung Dukuh along the Winongo, while the physical alterations and reinforcements to the river edge (gabions and rock-walls) are fairly new, the domestic elements that are already appearing along the riverside (potted plants, canopies, seating elements) illustrate the physical signs of the residents' care and ownership (Figure 7.3). The attitude of stewardship that these communities take is demonstrated through their continued maintenance of the riverside.

Use of the River and Riverside Spaces

Socioeconomic and Domestic Use

While in the urban sites the opportunities for adapted uses and transformations of the riverside spaces were limited, thus limiting opportunities for human activity; in the sub-urban sites the sociocultural significance of the river was still clearly evident. Daily uses of riverside areas included economic, domestic, and recreational functions. The Cikapundung, Winongo, and Code rivers all bore evidence of socio-economic and domestic activities occurring within and/or alongside the river. These included use of the river edges for fruiting or edible vegetation, fish and bird keeping, and the harvesting of sand for use in construction. In the case of sand harvesting, fine-grained naturally occurring sediment washed down the river was being incrementally collected (see also spatial aspects—expansions).

While in a natural system this might be viewed as destructive, within such a dense urban environment the harvesting of sediments can be viewed as a positive contribution to the maintenance of the functionality of the river channel.

Recreation and Access to the River

For discussion of recreational use, specifically, see "Open Space: Parks and Playgrounds" in the next section.

Spatial Aspects

Specific to riverine vernacular settlements, spatial aspects of landscape transformations included the re-orientation of settlements towards the river—including promoting the visibility of the river from the settlement, and the establishment of access ways, to facilitate access along the river and between the river and the city. Table 7.1 details the types of activities or transformations that these entail, and prevalence within neighborhoods included in the case study.

Access

Continuous Walkways along the River
Continuous walkways were observed to have been implemented along the river amongst four of the case study settlements which were located in urban and suburban areas of Yogyakarta and Bandung (Figure 7.4). These walkways were largely constructed from stone and concrete, and edged with walls. Except in the steep riverbank area of Tegal Panggung, houses and apartments were oriented toward the walkway—which served to provide them with improved access—and thus, to the river.

Facilitated At-Grade Access to the River
Facilitated at-grade access to the river was observed in only two of the case study settlements, the Cikapundung and the Winongo (Figure 7.5). In the former case no immediate benefit was visible, however, in the latter case the river access served to allow a nearby resident access to a foreshore formed by sedimentation along the river edge. This area was then planted with fruiting and edible vegetation species. Despite there only being two situations where river access was deliberately provided this was not the only

Figure 7.4 **Paths along rivers both provide access to the neighborhood and allow greater numbers of residents to come into proximity with the river (Photo by the Author, 2012)**

Figure 7.5 **Facilitated at-grade access to the river is less common among the case study settlements (Photo by the Author, 2012)**

instance where such behavior was observed, and this is described also in the section below titled "Informal foreshores."

Connections across the River (Bridges)
Pedestrian connections such as bridges between riverside neighborhoods were only observed in two of the case study settlements, the Cikapundung and the Winongo. While the sturdy steel and concrete bridge in downtown Bandung provided access for pedestrians and motorcyclists; the narrow steel and timber crossing between suburban riverside neighborhoods in Yogyakarta was also used for recreational activities such as fishing and socializing by men and boys.

Expansions

Informal Foreshores (Sedimentation)
Informal foreshores, forming from naturally occurring sedimentation along the river corridor, were observed to often be adapted by local residents for various socio-economic and domestic functions. Although facilitated access to the river was only observed at two case studies, amongst four of the settlements which were located in urban and suburban areas of Yogyakarta and Bandung residents were observed to have accessed the river corridor for a variety of purposes. All of these areas were seasonally exposed. Along the Cikapundung River a man was observed harvesting sand from a foreshore that had been colonized by grasses. Meanwhile along the Winongo River, adjacent to a staircase providing river access, a resident has cultivated a variety of edible and fruiting plants (such as *Carica papaya*) for domestic consumption. In Kampung Code on the Code River a larger foreshore area—evidently there for some time because of the permanence of its adaptation—has a rectangular concrete fishpond, and banana culms (*Musa sp.*) among others (Figure 7.6). In the neighborhood of Tegal Panggung ducks are accommodated in bamboo pens and roam free on a foreshore area, which is limited by the river edge and the river. These opportunistic and enterprising activities of locals (Ramírez-Lovering 2008) are vulnerable to city-scale maintenance regimes, such as dredging to ensure smooth flows during the wet season (Toumbourou 2009). However, these adaptive uses of foreshore areas provide tangible evidence of the sociocultural and economic value of the river within the everyday lives of riverside communities.

Figure 7.6 Adaptive uses of foreshore areas provide tangible evidence of the sociocultural and economic value of the river within the everyday lives of riverside communities (Photo by the Author, 2012)

Open Space: Parks and Playgrounds

Open spaces along the riverside were observed in various forms. Amongst the three case studies from Yogyakarta, open spaces were found along the river, implemented with varying success. In the case of the suburban case study along the Winongo River, a neighborhood playground has been outfitted with children's play equipment, brightly painted planters, and paved open space (Figure 7.7). The suburban setting facilitates a green, vegetated outlook.

Comparatively, open spaces along the Code River face greater limitations as its banks are significantly steeper and the urban area is more densely settled and less permeable due to the prevalence of hard surfaces such as concrete and asphalt. Despite this, an open space directly at the river's edge in Kampung Code is lively and well-used. It buffers domestic activities (such as clothes drying), which are constrained within the dense neighborhood, and provides space for children to play. Additionally, activities such as fish and bird keeping, whose profits provide additional income as well as opportunity for use within the home, take place adjacent to and within the space. In the neighborhood of Tegal Panggung, steep terrain limits open

Figure 7.7 Open spaces within settlements—particularly with provision for children, such as playgrounds—are rare (Photo by the Author, 2012)

space provision with only walkways accommodated. On the opposite side of the river at Suryatmajan, a public housing block (*rumah susun*) has been configured with open space both below the block and directly adjacent to the river. Compared to the apartments at Tegal Panggung, this space is relatively unused by the residents and less lively as a result. It appears that ground-floor activation—such as the residential ground floor at Tegal Panggung—may be important in facilitating community stewardship of a neighborhood.

Along the Cikapundung River, open spaces are rarer and were mostly observed within the urban fabric, away from the river's edge. In the suburban area, a densely vegetated riverside park is equipped with a small pavilion (Figure 7.8). Community groups are involved in revegetation and education programs within the surrounding area. Contrastingly, along the Bengawan Solo River near Surakarta the resettlement of a neighborhood in compliance with Regulation 38/2011 has allowed for a park and playground to be located within the floodplain (inside the embankment), along with the urban forest described earlier. The significance of spatial constraints on the design of open spaces within riverside settlements is clearly evident when comparing these case studies.

Figure 7.8 Environmentally-focused open spaces are found as a result of the efforts of community groups and innovative planning (Photo by the Author, 2012)

Limits/Boundaries

Walls and Embankments

Amongst the urban and sub-urban case studies flood protection was observed to be facilitated largely through the use of walls and embankments. While walls were observed in varying heights and widths in all of the urban and sub-urban case study settlements, only the Bengawan Solo was observed to have necessitated the implementation of an embankment (Figure 7.9). This is perceived to mainly be due to the dense settlement of the other case study neighborhoods and their often steep terrain. In Yogyakarta walls have been used in riverside neighborhoods since the late 1980s for protection from flooding, however the Merapi eruption in 2010—in which lahar flowed along the rivers flooding the city—highlighted that higher water levels were not the only threat in this region. The wider walls implemented along the Code River in the Kampung Code neighborhood are adapted by residents and are used for domestic activities such as drying fish and clothing, and recreational activities such as socializing and fishing (Figure 7.10). In comparison, the narrower walls implemented at the Cikapundung and Winongo riverside neighborhoods were observed to be

Figures 7.9 and 7.10 Embankments along the Bengawan Solo River and gabion walls along the Code River provide protection from flood (Photos by the Author, 2012)

adapted for pot plants, and the narrow wall along the Code River in Tegal Panggung was observed to only provide a delineation of and protection against the steep incline down to the river. In the case of the Bengawan Solo, riverside settlements have been relocated and an embankment protects the city from flood. A road along the top of the embankment facilitated access along the river.

Conclusion and Outlook

Within this research, investigation methods from environment and behavior research have been used to observe physical traces of human activities and changes made to surroundings. Changes made to surroundings for the purpose of the physical improvement of the riverside, and as a result or reaction to the improvement of the waterfront, were of particular interest. The behavior of communities living along the river's edge was observed to be affected by physical traces such as barriers, walls, orientation, and objects, and as such, these elements have informed the structuring of our results. Generally, the measures implemented at the sites selected include physical improvements, community participation, facilities, housing improvement, tenure regularization, and development programs, which were found to be consistent with the UN-Habitat's principles of participatory slum upgrading (UN-Habitat 2012) (Table 7.2). While this could certainly be perceived as a baseline for transformation within densely settled areas, like the KIP, the UN-Habitat principles do not focus on changes to particular kinds of informally settled environments, such as riversides. The following conclusions were observed as common to positive examples of riverine landscape transformation when undertaking river improvement in Indonesia:

First, transformations are often triggered from the "ground-up," with communities and/or NGOs driving improvements to the physical environment. In cases such as the Code River in Yogyakarta, changes to the river environment were proposed and implemented by communities and NGOs. This is significant as residents can contribute site-specific knowledge based on their experiences living in riverside neighborhoods (Vollmer et al. 2015; Prescott and Ninsalam 2016; Padawangi et al. 2016), which may result in more self-sustaining improvements (Table 7.3).

Table 7.2 Summary of actions taken to improve riverside settlements broken down into sociocultural and spatial factors. Actions that have been asterisked correspond to some degree to the UN-Habitat's principles of "participatory slum upgrading".

Sociocultural Transformations	
Empowerment	Community participation* Tenure regularization* Development programs*
Respect for Environment	Activities/programs Groups
Spatial Transformations	
Physical Improvements*	Orientation (to river) Access (to river and city) Sanitation, streets, and drainage* Greening (vegetation)
Amenities and Facilities*	Public open spaces Libraries, schools, community halls

Table 7.3 Summary of range of parties involved in selected case studies, and actions taken to improve riverside settlements broken down into sociocultural and spatial factors

		Case study				
Variables		Cikapundung River, Bandung	Sungai Winongo, Yogyakarta	Kali Code (Kampung Code), Yogyakarta	Kali Code (Tegal Panggung), Yogyakarta	Bengawan Solo River, Surakarta
Involvement	Neighborhood	x	x	x		
	NGO	x	x	x		
	Institution					
	Regional/municipal government		x		x	x
Sociocultural	Community empowerment			x		x
	Respect for environment	x	x	x		x
Spatial	Orientation	x	x	x	x	
	Access / preserve edge / set-back	x	x	x	x	x
	Infrastructure / facilities			x	x	

Second, community involvement was an important sociocultural factor within the case study sites. Amongst these case studies various opportunities for ongoing engagement with the river environment were observed, such as environmental programs, and served to educate and involve residents in changes within the river corridor. Additionally, the siting of cultural activities such as festivals and community celebrations along (and within) the river added to a sense of ownership and responsibility toward the river. These all contributed to a sense of, and indeed evidence of, environmental stewardship being present within neighborhoods. Community involvement was also observed to contribute to ongoing maintenance of riverine landscapes, with neighborhoods engaging in upkeep of vegetation buffers, and sediment harvesting—all of which can help to sustain river flow.

Third, positive examples of river improvement in Indonesia, such as those described here, reveal the need for physical connections between neighborhoods and rivers. Amongst these case studies, lively and cared-for riversides, like that of the Winongo and Kampung Code in Yogyakarta, illustrate the role of spatial orientation, accessibility, and scaling of spaces in establishing connections between neighborhoods and rivers. Obstacles such as high walls and embankments were physical barriers which limit interaction between neighborhoods and the river (see also Setiawan 2006). The following were seen to be key spatial measures in the studied neighborhoods: re-orientating neighborhoods to face the river, ground-floor activation, providing pedestrian access along and to the river, and providing spaces of adequate size for social, recreational, and other domestic or economic activities (whereby they do not impinge on the condition of the river themselves). As a result, these supported an ongoing relationship between communities and the river, providing an opportunity for residents to interact directly with the river environment (see also Setiawan 2006). Additionally, this physical proximity ensured that residents were given a means to observe the temporal cycles of the river, which in turn supports resiliency by making them aware of seasonal and ecological changes. Retaining sightlines, and making channel dynamics visible, has been discussed to positively influence the ability of communities to monitor changes of rivers, which may lead to increased safety (Prominski et al. 2012). Interestingly, although government policy requires vehicular access for river channel maintenance and assessment (Republik Indonesia 2011), none of these instances observed within the case studies exhibited signs of this directly along the river's edge. While this could be seen as increasing risk during emergency events, vehicles were still able to access all neighborhoods to some degree and riverside paths were wide enough for small vehicles such as motorcycles and carts. Other studies have observed that communities can be involved in the assessment of river conditions, which contributes to community resilience during disturbances (Vollmer and Grêt-Regamey 2013).

Fourth, resilience or tolerance of designs and communities to ongoing environmental disturbance (as illustrated by case studies in Surakarta and Yogyakarta) was similarly important. Based on their physiology, the rivers suffer sedimentation (evidenced by the foreshores that have formed along their banks) as well as flooding. From the Yogyakarta case study, it was observed that volcanic eruptions were also a potential hazard. Therefore, some degree of physical separation can be seen as necessary to protect neighborhoods. However, at the same time it is important that sociocultural aspects are seen by government authorities and external stakeholders like donor foundations as valuable, in order to maintain connection between the communities and the river. This reveals the challenge of separating—for the purpose of protection—while still maintaining visible and physical connections between neighborhoods and the river. As such, benefits can be seen in implementing stepped/transitional designs, facilitating river access through selective spatial expansion, and providing submersible or tolerant infrastructures and structures (Prominski et al. 2012) such as walkways and building on stilts, as in vernacular Indonesian architecture. Existing stepped typologies were observed to be likely more a result of neighborhoods working with existing, often steep, terrain. All of these approaches may contribute to neighborhood resilience during events such as floods.

The relationship between society, culture, and landscape is well established (Meyer 2008). The case studies explored within this chapter illustrate the potential of projects centered on neighborhoods and the local scale to have positive environmental impacts. Through using a sociocultural and spatial framework of analysis a range of improvement measures has been revealed, while the comparative approach showed benefits, or landscape services, of particular approaches. Thus a comparative approach at the local landscape scale is demonstrated to be a potentially useful tool in understanding the range of measures, their spatial constraints, and provides a basis for future local-river specific planning. A key result of this study has been the generation of a more comprehensive and combinatory inventory of strategic design tools for designers and planners. Table 7.3 serves as a base for considerations during the research, thus leading to a wider methodological reflection useful in practice. Applying such a spatially explicit methodology, based on locally specific spatial typologies, provides useful insights on how particular spatial measures, or typologies, contribute to the behaviors of those living within riverine landscapes and what aspects should be considered for the design of durable improvements. Based on this study, sociocultural factors (such as education, involvement in changes, and consideration of the social, cultural, and economic needs of residents and communities) along with physical factors have been reinforced to be vital to sustained improvement of river edges. Since communities have a strong impact on the extent that urban riparian landscapes can be revitalized, it is important that the created or modified ecosystem matches community values, taking into account both an ecological habitat perspective, as well as residents' preferences for open space, shade trees, and garden space (Findlay and Taylor 2006; Vollmer et al. 2015). In so doing, the reconfiguration, including greening and habitation, of river edges could contribute to a range of other co-benefits, as well as reinforce the Indonesian government's commitment to sustainability.

Acknowledgments

I would like to acknowledge the hospitality of communities during my fieldwork in Indonesia. I am grateful to Dr. Ramalis Sobandi of Tunas Nusa Foundation for facilitating interactions and site visits with communities in Surakarta and Bandung, and to Dr. Mahditia Paramita and staff at the Housing Resource Centre for providing access to reports, and facilitating interactions and site visits with communities in Yogyakarta. The hospitality and rich discussions with these women and their colleagues was invaluable to the development of this research. Finally, I would like to thank the Singapore National Research Foundation and Singapore-ETH Centre Future Cities Laboratory for providing the financial support to carry out the research and associated fieldwork.

References

Abeyasekere, Susan. 1987. *Jakarta: A History*. Singapore: Oxford University Press.

Adi, Ganug Nugroho. 2012. "Reviving the Solo River's Heyday." *The Jakarta Post*, 28 November 2012, accessed 12 June 2019.

Antara News. 2017. "Indonesia Hit by More Floods in 2017." 26 October 2017, online. Available at https://en.antaranews.com/news/113220/indonesia-hit-by-more-floods-in-2017, accessed 12 June 2019.

Antrop, Marc. 2005. "Why Landscapes of the Past Are Important for the Future." *Landscape and Urban Planning*, 70: 21–34.

Arifin, Hadi Susilo, Regan Leonardus Kaswanto, and Nobukazu Nakagoshi. 2014. "Low Carbon Society through Pekarangan, Traditional Agroforestry Practices in Java, Indonesia." In *Designing Low Carbon Societies in Landscapes*, edited by Nobukazu Nakagoshi and Jhonamie A. Mabuhay, 129–143. Tokyo: Springer.

Arifin, Hadi Susilo, and Nobukazu Nakagoshi. 2011. "Landscape Ecology and Urban Biodiversity in Tropical Indonesian Cities." *Landscape and Ecological Engineering*, 7(1): 33–43.

Asian Development Bank. 2007. *Proposed Loan Republic of Indonesia: West Jakarta Water Supply Development Project*. Extended Annual Review Report. Mandaluyong City, Philippines: Asian Development Bank. Retrieved 13 June 2019.

Asian Development Bank. 2012. *The Neighborhood Upgrading and Shelter Sector Project in Indonesia: Sharing Knowledge on Community-Driven Development*, 36. Mandaluyong City, Philippines: Asian Development Bank. Retrieved 13 June 2019.

Benedict, Mark A., and Edward T. McMahon. 2002. "Green Infrastructure: Smart Conservation for the 21st Century." *Renewable Resources Journal*, 20(3): 12–17.

De Meulder, Bruno, and Kelly Shannon. 2013. *Water Urbanisms East: Emerging Practices and Old Traditions*. Zurich: Park Books.

The Economist. 2013. "China and the Environment: The East Is Grey." *The Economist*, 10 August 2013, 17–20.

Farida, Winia, Lufiandi, and Setiawan Wangsaatmadja. 2007. "The Sanitation Facilities Improvement Based in Social and Cultural Consideration in Kelurahan Batununggal, Bandung City, Indonesia." In *Proceedings: International Forum on Water Governance in Asia*, 14–15 March 2007. Bangkok: Water Environment Partnership in Asia (WEPA).

Feldman, Roberta M., and Lynne Westphal. 2000. "An Agenda for Community Design and Planning: Participation and Empowerment in Practice." In *Sustaining Human Settlement: A Challenge for the New Millennium*, edited by Roderick J. Lawrence, 105–139. North Shields, CA: Urban International Press.

Findlay, Sophia J., and Mark P. Taylor. 2006. "Why Rehabilitate Urban River Systems?" *Area*, 38(3): 312–325.

Francis, Mark. 2001. "A Case Study Method for Landscape Architecture." *Landscape Journal*, 20(1): 15–29.

Gill, Ronald. 1993. "The Architectural Urban Heritage of Jakarta: A Case Study for Planning Indonesia for Cultural Tourism in Cities in Indonesia." In *Cultural Tourism, International Scientific Committee on Cultural Tourism, Sri Lanka*, 70–82. Columbo: National Committee of ICOMOS and the International Council on Monuments and Sites.

Greco, Steven E., and Eric W. Larsen. 2014. "Ecological Design of Multifunctional Open Channels for Flood Control and Conservation Planning." *Landscape and Urban Planning*, 131: 14–26.

Hermann, Anna, Michael Kuttner, Christa Hainz-Renetzeder, Éva Konkoly-Gyuró, Ágnes Tirászi, Christiane Brandenburg, Brigitte Allex, Karen Ziener, and Thomas Wrbka. 2014. "Assessment Framework for Landscape Services in European Cultural Landscapes: An Austrian Hungarian Case Study." *Ecological Indicators*, 37: 29–240.

Kumorotomo, Wahyudi, Muhadjir Darwin, and Faturochman. 1995. "The Implementation of Slum and Squatter Improvement Programs in the River Basins of Yogyakarta." *Populasi*, 6(2). http://dx.doi.org/10.22146/jp.11457.

Lucas, Anton. 1998. "River Pollution and Political Action in Indonesia." In *The Politics of Environment in Southeast Asia: Resources and Resistance*, edited by Philip Hirsch and Carol Warren, 181–209. London: Routledge.

Lucas, Anton, and Arief W. Djati. 2007. "The Politics of Environmental and Water Pollution in East Java." In *A World of Water*, edited by Peter Boomgaard, 321–351. Leiden: KITLV Press.

Meyer, Elizabeth K. 2008. "Sustaining Beauty: The Performance of Appearance. A Manifesto in Three Parts." *Journal of Landscape Architecture*, 3(1): 6–23.

Morgan, Sandra, and Ann Bookman. 1988. "Rethinking Women and Politics: An Introductory Essay." In *Women and the Politics of Empowerment*, edited by Ann Bookman and Sandra Morgen, 3–29. Philadelphia, PA: Temple University Press.

Nakamura, Keigo, Klement Tockner, and Kunihiko Amano. 2006. "River and Wetland Restoration: Lessons from Japan." *Bioscience*, 56(5): 419–429.

Nassauer, Joan Iverson. 1995. "Messy Ecosystems, Orderly Frames." *Landscape Journal*, 14(2): 161–170.

Nassauer, Joan Iverson, and Paul Opdam. 2008. "Design in Science: Extending the Landscape Ecology Paradigm." *Landscape Ecology*, 23(6): 633–644.

Padawangi, Rita, Etienne Turpin, Michaela F. Prescott, Ivana Lee, and Ariel Shepherd. 2016. "Mapping an Alternative Community River: The Case of the Ciliwung." *Sustainable Cities and Society*, 20: 147–157.

Prescott, Michaela F., and Yazid Ninsalam. 2016. "The Synthesis of Environmental and Socio-cultural Information in the Ecological Design of Urban Riverine Landscapes." *Sustainable Cities and Society*, 20: 222–236.

Prescott, Michaela F., and Yazid Ninsalam. 2020. "River Communities in Jakarta: Pedagogies in Environmental Stewardship." *Journal of Environmental Studies*, 66 (Citizen Participation Pedagogy): 70–95.

Prihandrijanti, Maria, and Mayrina Firdayati. 2011. "Current Situation and Considerations of Domestic Waste-Water Treatment Systems for Big Cities in Indonesia (Case Study: Surabaya and Bandung)." *Journal of Water Sustainability*, 1(2): 97–104.

Prominski, Martin, Antje Stokman, Daniel Stimberg, Hinnerk Voermanek, and Susanne Zeller. 2012. *River. Space. Design: Planning Strategies, Methods and Projects for Urban Rivers*. Basel: Birkhauser.

Raharjo, Wiryono. 2010. *Speculative Settlements: Built Form/Tenure Ambiguity in Kampung Development*. PhD Thesis. Melbourne School of Design.

Ramírez-Lovering, Diego. 2008. *Opportunistic Urbanism*. Melbourne: RMIT Publishing.

Ramírez-Lovering, Diego, Michaela F. Prescott, and H. Kamalipour. 2018. "RISE: A Case Study for Design Research in Informal Settlement Revitalisation. Interdisciplinary Design Research in Informal Settlements." In *Proceedings of the 1st Annual Design Research Conference* (*ADR18*), edited by Duncan W. Maxwell, 461–478. Sydney: The University of Sydney.

Reerink, Gustaf. 2011. *Tenure Security for Indonesia's Urban Poor: A Socio-legal Study on Land, Decentralisation, and the Rule of Law in Bandung*. Leiden: Leiden University Press.

Republik Indonesia. 2011. Peraturan Pemerintah 38/2011. 2011. Tentang Sungai.

Sakijege, Tumpale, Junun Sartohadi, Muh A. Marfai, Gabriel R. Kassenga, and Samson E. Kasala. 2014. "Assessment of Adaptation Strategies to Flooding: A Comparative Study between Informal Settlements of Keko Machungwa in Dar es Salaam, Tanzania and Sangkrah in Surakarta, Indonesia." *Jàmbá: Journal of Disaster Risk Studies*, 6(1): 131.

Scheuer, Dominik. 2014. Interview with Office of Community Empowerment, Surakarta.

Seftyono, Cahyo. 2011. "Local Community in Valuing Ecosystem Services: Warga Kampung Code's Perspective on Kali Code Existence." In *Proceedings of Annual Indonesian Scholars Conference in Taiwan, 2010* (*2011*). Tainan. Available at http://ssrn.com/abstract=1763326.

Seftyono, Cahyo. 2012. "March. Kali Code: From Merapi to Sustainable Society." In *International Conference on Sustainable Innovation 2012*, 19–20. Available at http://ssrn.com/abstract=2020132.

Setiawan, Bakti. 2006. "Ruang bermain untuk anak di kampung kota: Studi persepsi lingkungan, seting, dan perilakuanak di Kampung Code Utara, Yogyakarta [Space for Children in Urban Kampung: Study on the Environmental Perception, Setting, and Behavior of Urban Children]." *Jurnal Manusia dan Lingkungan*, 3(2): 60–70.

SMERU Research Institute. 2011. *Addressing Urban Poverty, Increasing Learning on Urban Poverty Reduction in Kota Surakarta and Kota Makassar, Indonesia*. Draft research report supported by the Ford Foundation, New York.

Sukardi, Sarwono, Bambang Warsito, Hananto Kisworo, and Sukiyoto. 2013. *River Management in Indonesia*. Jakarta: Directorate General of Water Resources, Ministry of Public Works and Settlements. Available at https://vdocuments.mx/20130830-river-management-in-indonesia-eng-vers.html.

Tahir, Akino, Sachihiko Harashina, and Mitsuo Yoshida. 2011. "Community Involvement in Solid Waste Management in Indonesia." In *Proceedings of the 2011 World Congress of International Solid Waste Association* (*ISWA*), 1: 83.

Takeuchi, Kuniyoshi, A.W. Jayawardena, and Yutaka Takahasi (eds.). 1995. *Catalogue of Rivers for Southeast Asia and the Pacific, Vol. 1.* UNESCO-IHP Regional Steering Committee for Southeast Asia and the Pacific, October.

Termorshuizen, Jolande W., and Paul Opdam. 2009. "Landscape Services as a Bridge between Landscape Ecology and Sustainable Development." *Landscape Ecology,* 24(8): 1037–1052.

Toumbourou, Tessa. 2009. "Down by the Riverside: Kali Code." *Inside Indonesia,* 95(Jan.–Mar.). Available at http://www.insideindonesia.org/down-by-the-riverside-kali-code, accessed 16 June 2019.

Tunas, Devisari, and Andrea Peresthu. 2010. "The Self-Help Housing in Indonesia: The Only Option for the Poor?" *Habitat International,* 34(3): 315–322.

UN-Habitat. 2012. *Streets as Tools for Urban Transformation in Slums: A Street-Led Approach to Citywide Slum Upgrading.* Nairobi: UN Habitat. Available at https://unhabitat.org/streets-as-tools-for-urban-transformation-in-slums.

Vollmer, Derek, and Adrienne Grêt-Regamey. 2013. "Rivers as Municipal Infrastructure: Demand for Environmental Services in Informal Settlements along an Indonesian River." *Global Environmental Change,* 23(6): 1542–1555.

Vollmer, Derek, Michaela F. Prescott, Rita Padawangi, Christophe Girot, and Adrienne Grêt-Regamey. 2015. "Understanding the Value of Urban Riparian Corridors: Considerations in Planning for Cultural Services along an Indonesian River." *Landscape and Urban Planning,* 138: 144–154.

Windarti, Yulia. 2016. "Local Government Attitudes toward Sustainable Tourism Development: Case of Bandung City, Indonesia." *International Journal of Social Science and Humanity,* 6–7: 551–556. https://doi.org/10.7763/IJSSH.2016.V6.709.

Yossi, Briandara, and Edsel E. Sajor. 2006. "Development of Riverside Kampungs and Management of Rivers in Yogyakarta, Indonesia: Issues of Policy Coherence and Relevance of Socio-economic Characteristics of River Bank Communities." In *Proceedings: Regional Conference on Urban Water and Sanitation in Southeast Asian Cities,* 297–322. Pathumthani, Thailand: Asian Institute of Technology.

Zeisel, John. 2006. *Inquiry by design: Environment/Behavior/Neuroscience in Architecture, Interiors, Landscape, and Planning.* New York, NY: W.W. Norton & Co.

Biography

Michaela F. Prescott's research focuses on the socio-spatial dimensions of landscapes and urban environments, and the evolving relation of landscape, infrastructure, and urbanization. Trained as a landscape architect, she undertakes action-research projects related to urbanization of water systems, infrastructure's impact on neighborhoods and communities, and delivery and sustainability of community-based projects.

8 Recovering the Stream

River Access as a Catalyst for Political Mobilization in Suwon City, South Korea

Youngah Guahk

Abstract

This chapter looks at the contestation over river access in an industrial city of South Korea. In the city of Suwon, a small stream had been covered with concrete in the context of rapid economic development during the 1990s—a decision taken in order to facilitate additional road construction. A decision that was originally uncontested, indeed welcomed by local business, did eventually generate an environmental and societal backlash, leading to the restoration of access to the stream. Beyond this, the experience had wide-ranging repercussions for local politics and the identity of the city. This chapter discusses the way in which access to the river became a key issue in local politics and ultimately led to the transformation of Suwon into a recognized eco-city.

Keywords: river access, South Korea, contestation, environmentalism, political mobilization, eco-city

Introduction

According to the OECD (Kamal-Chaoui et al. 2011), more than half of the world's total population live in cities and this number is only going to increase over time. Scholars have estimated that this urban population is likely to grow, especially in developing countries (Wong and Yuen 2011; Hu Mei-Chih et al. 2016). In Asia alone, more than one billion people are expected to move to cities in the next two decades. By 2011, this urban explosion included 11 mega cities in Asia, such as Beijing, Shanghai, Tokyo, Delhi, Jakarta, and Seoul, each of which had a population of more than

Padawangi, Rita, Paul Rabé and Adrian Perkasa, *River Cities in Asia*.
Taylor & Francis Group, 2022
DOI: 10.5117/9789463721851_CH08

10 million inhabitants (Wong and Yeun 2011, 1). Cities play a major role in economic growth and job creation. At the same time, cities also contribute to massive urbanization and migration problems such as air and water pollution, acid rain, housing and water shortages, as well as other issues (Lye and Chen 2010).

In reality, innovative cities are essential to address the challenges of urbanization, economic growth, and development in the East Asia region, but overall, as the population of cities quickly expands, this also increases the pressure on governments to manage their growth (Johnson 2007). One of the specific concepts that attempts to link urban growth, sustainable development, and innovation in East Asia is that of the "eco-city," a development that can be observed in all three of the major economies of East Asia (Kamal-Chaoui et al. 2011; Hu Mei-Chih et al. 2016; Wong and Yuen 2011).

The concept of the eco-city is closely related to the idea of sustainable development (Leznicki and Lewandowska 2014, 93). Wong and Yuen have argued that the eco-city concept guides an "ecological approach to urban design, management and towards a new way of lifestyle" (2011, 3). It is notable that the concept of eco-city is often linked in discourses with terms such as "green," "agriculture," "ecology," "natural environment," "food producing," and others (Kenworthy 2006; Leznicki and Lewandowska 2014; Roseland 1997). While some international bodies award some cities with the label "eco-city," there is no universally accepted international standard for what qualifies as an eco-city.

A particular aspect of the "eco-city" concept is the relationship between a city and its water resources. As a city grows, its residents also require greater access to water for drinking, living, and sanitation. Rivers are also of important ecological, social, and cultural value and "eco-city" projects are therefore frequently concerned with the environmental protection of rivers. However, urbanization and economic development also make claims on the commercial use of rivers, be it through shipping or as convenient locations for industrial production, energy generation, or the like. This is one particular dimension of the familiar contestation between ecological concerns and economic interests.

Yet, the presence of a river in an urbanized environment also has the potential to contribute to another contestation over the use of urban space. If, as in the case discussed in this chapter, the river—or rather the stream—is too small to be utilized by itself as a transport route or for other commercial use, economic interests may push for the abandonment of the river altogether. In such cases, from a narrow (and superficial) commercial perspective, the river might be seen as a "waste of space" that can be considered for transformation into a different kind of space, a scenario that is particularly pertinent in the context of countries such as South Korea that have undergone a process of rapid industrialization, often with little regard to the possibility of environmental damage.

This chapter addresses a specific research agenda related to this problematic, namely the impact that contestation over access to the riverfront can have on urban politics and local society. It does so by examining a particular experience in South Korea, namely in the city of Suwon, where access to a stream and the usage of urban space constituted the source of political conflict and the foundation for a process of urban renewal.

In Suwon, an industrial city near the capital, Seoul, the local government decided to cover up the Suwoncheon stream in 1988, in order to provide more space for roads, parking, and other urban developments. The issue here was not so much about pollution or neglect of the city's water resources, but rather the active removal of access to water and thereby fundamentally changing the cityscape. This policy also had implications for the city's heritage, given that the stream was connected to a cultural heritage site. Consequently, the concreting over of the stream could have compromised the historical value and threatened its potential candidacy as a UNESCO World Heritage site.

The initial decision to cover up this stream, and the subsequent opposition to it, had major repercussions for the city's future development: it gave rise to an environmentalist movement that mobilized a large part of the local population in favor of stopping the development and ultimately returning the stream to its former status. It also changed the nature of local politics, elevating the leader of the environmental movement to the position of mayor and putting the city on a trajectory towards becoming an eco-city.

As such, the experience makes Suwon a very particular example of a river city. A small stream in a growing urban center was regarded, in the context of rapid industrialization, as an obstruction that could be disposed of. City authorities began covering up the stream in order to utilize the space commercially. Yet these actions mobilized activists and created a growing environmentalist movement that ultimately took control over local politics and changed the outlook of the city to that of an "eco-city." In the context of these developments, it became evident how central even a small river can be to the identity of a city. What mattered in the case of Suwon was not only the contestation between ecological concerns and economic interests, but also the cultural, historical and symbolic significance of the river to the inhabitants of the city.

The purpose of this chapter is to chart the process of transformation in this South Korean city, with particular emphasis on the changing nature of local politics that was triggered by the dispute over the protection of the river environment. The chapter begins by outlining the structural conditions of the city and the nature of the environmental "crisis" created by the decision to cover up the stream. It then moves to a discussion of the local politics surrounding this issue, including the role played by local decision-makers, NGOs, and the private sector, before examining the transformation that the city has undergone in the past two decades. Ultimately, this process involved fundamental political change, namely the election of a former environmental activist as a mayor, and set the city on the path towards becoming an eco-city. The chapter concludes by highlighting the impact that decisions concerning the development of water spaces can have in an industrial city in South Korea.

The study was done through a qualitative case study of Suwon city in South Korea. The data were collected via archival research and "in-depth" interviews with local elites who had been deeply involved with the process. Dexter defined "the elite" as the more "influential, the prominent, and the well-informed than the rank-and-file of a population" (1970, 19). Therefore, the local elite in Suwon is made up of mainly civil servants including the city mayor as well as members of local councils. Furthermore, leading members of NGOs and representatives of local think-tanks were also interviewed in order to triangulate the findings from both written and oral sources of information as the literature suggested (Tight 2017, 4).

Moreover, process-tracing was used to analyze how environmental contestation created an opportunity for a new eco-city development in an ordinary industrial city. The process-tracing method is a tool of qualitative analysis (Collier 2011, 823) and helps to reveal the causal chain or mechanism regarding certain processes or policy fields. This method analyzes evidence on processes and conjunctures of events to develop or test hypotheses about causal mechanisms that may explain the case (Bennett and Jeffrey 2014, 7).

Suwon City's Geographical and Cultural Features

Suwon city is a lower-tiered government with 1.2 million residents, located in the center of Gyeonggi province, 40.7 kilometers south of South Korea's capital city, Seoul. Suwon is also the capital of the province and shares a boundary with many cities in Gyeonggi province.

The Seoul-Gyeonggi Metropolitan Area is the most populated area in South Korea with more than 25 million inhabitants and it includes major regional governments such as Seoul Capital city, Incheon Metropolitan city and Gyeonggi province. Because of this, the region has been enlarged tremendously since the 1980s with a number of new towns (Shin Doshi) developed (Lee 2012). While Seoul itself also underwent a rapid process of urbanization and industrialization—and experienced its own urban regeneration project around the well-known Cheonggyecheon redevelopment (Lee and Anderson 2013)—this particular experience in the city of Suwon is distinct from the wider regional developments (Kwon 2008).

Suwon city is known as a gate city, with major traffic roads passing through it (Suwonshi 2015a, 57). For example, the Gyeongbu Motorway connects Seoul to Busan metropolitan city in the Southeast of the Korean peninsula. Furthermore, the Yongin-Seoul Motorway, Kyungin Industrial Road, and Gwacheon-Uiwang Road all pass through Suwon city territory and thus Suwon city is considered to be one of Korea's major traffic points (Suwonshi 2015a, 59).

Having access to all those motorways might be necessary and helpful for the economy of Suwon city. However, being subject to the high volume of traffic also raises concerns about air quality in the Suwon area. According to a Suwon Environment White Paper, in 2009, over 60 percent of the air pollution in Suwon city was caused by auto emissions (Suwonshi 2015b, 105).

Seoul Metro Line 1 allows local residents to travel to Seoul, and in 2003 this line was extended to other cities, further south in Gyeonggi province. In addition, the Gyeongbu railroad linking Seoul to Busan stops at Suwon city (Suwonshi 2015a, 59). Moreover, several different public bus lines have been established in and out of Suwon city.

Due to its central location and its connectivity to the capital city and other large cities in Gyeonggi province, Suwon forms part of the "conurbation" in the Seoul-Gyeonggi metropolis (Suwonshi 2015a, 58). This is similar to an "urban corridor," with several cities clustering in a corridor surrounding one mega city. In this case, Suwon and other big cities with around one million inhabitants—such as Goyang city, Seongnam city, Yongin and Ansan city—surround Seoul (Suwonshi 2015a, 59).

Suwon's geography has the following distinctive characteristics: first, the land is elevated on the northeast side and there are several mountains in the area. Second, no big river flows through Suwon city. Instead, four main water streams (*cheon* in Korean) pass through the city, namely, Hwanggujicheon, Woncheonricheon, Suwoncheon, and Seohocheon. Third, Suwon city is home to seven large water reservoirs: Gwanggyo, Woncheon, Shindae, Seoho, Mansoeggoe, Ilwol, and Wangseong. These streams and reservoirs add green land to the city and also play a role as recreational areas (Suwonshi 2015b, 22).

With its proximity to Seoul and its highway connectivity to other regions, Suwon is recognized as an industrial city. In 2010, a Suwon government document specified that there are 47 large companies and conglomerates in Suwon city and the surrounding area (NHS Trust 2010, 17). Four out of the 47 large companies are located in the Suwon city territory (Suwonshi 2015a).

Despite Suwon city being known as an industrial city, the city itself has a long and rich history, evidenced by the numerous historical monuments in the city center. The most famous one is the Hwaseong Fortress, which in the eighteenth century was a fortified wall surrounding the entire city. The fortress was built for defensive purposes to protect the tomb of King Jeongjo's father, Crown Prince Jangheon of the Joseon Dynasty (UNESCO 1997).

Construction on the original fortress began in 1794 and used the then "modern" technology of the period. The fortress wall has a length of 5.7 kilometers and covers an area of 130 hectares, based on the topography of the land itself. The Suwoncheon, the main water stream in the city, flows through the Hwaseong Fortress. UNESCO recognized the unique status of Hwaseong Fortress and designated it as a UNESCO World Heritage site in December 1997 (UNESCO 1997). The fortress is a symbol of Suwon city and is often used in the city's cultural and historical promotional activities.

Water Crisis in Suwon City: The Covering of Suwoncheon

Korea achieved fast economic growth in the 1970s to 1980s and as a consequence triggered internal migration by job seekers to the Seoul-Gyeonggi Metropolitan Area as well as to other large industrial cities in the country. This internal migration brought with it uncontrolled urbanization processes, causing environmental problems such as air, land, and water pollution in these big cities. Suwon experienced a similar development pattern and local residents encountered particular environmental problems in the mid-1990s, in particular the pollution of the Suwoncheon stream.

The Suwoncheon is a small stream passing through the center of the city. It is 16 kilometers long and has approximately 25 kilometers of basin area, which is a relatively small area. However, there are a number of old monuments and also many commercial buildings around the Suwoncheon due to its prime location in the city center. The particular historical monuments are the detached Summer Palace, Suwon Haenggung, and the Hwaseong Fortress, which was built around part of the Suwoncheon and includes a bridge gate (Hwahongmun) crossing the stream. In addition, major traffic roads cross over the Suwoncheon.

Five major traditional markets are also located in the Suwon city center near the Suwoncheon. These markets are called Nammun-sijang, Jidong-sijang, Paldalmun-sijang, Yeongdong-sijang, and Motgoljonghab-sijang. The market vendors sell everything from textile products to fresh foods. The presence of traditional markets means heavy traffic during the day and the need for plenty of parking spaces. Moreover, each market day sees the accumulation of large quantities of rubbish from the market.

In 1988, the Suwon local administrative government decided to cover the Suwoncheon, partially based on a presidential campaign commitment by then President Roh Tae-woo. Two main reasons were given for the decision: first, to solve the traffic congestion problem in the city center, and second, to increase parking space to attract more customers to the city center and to rejuvenate the surrounding commercial area, particularly the traditional markets. The covering construction officially began in 1991. Originally, 94 percent of the citizens agreed with the plan, according to an opinion poll conducted at the time. The 790 meters of the first phase of the construction, between Jidonggyo (Jidong Bridge) and Maegyo (Mae Bridge), was completed in 1994 (Kim 2002, 48).

Originally, the construction was planned to connect Maehyanggyo (Maehyang Bridge) to Maegyo for 1270 meters (Yoem 2012). However, the distance covered was just 480 meters, shortened due to conflict with the restoration plan for the historical monument Namsumun (South water gate) (Yoem 2012). The second phase of construction began in March in 1995 (Kim 2002, 49).

An action group also made an investigation into the environmental impact of the Suwoncheon construction and evidence showed that the covering had a negative impact on the water quality of the Suwoncheon before and after the completion of the first phase of construction. In 1995 environmental NGOs and local residents joined forces to investigate and it became clear that covering the stream severely affected its basin area. The water became polluted and within a year, without sunlight under the concrete cover, the flora died (Interviewee 3; Kim 2002, 48). In May 1996, construction was halted after 30 percent of the second phase had been completed.

However, the covered concrete parts between Jidonggyo and Maegyo remained for another decade and during that period the covered parts were used as parking space for visitors to the traditional markets and city center. Without sunlight, water under the covered part lost its self-purification capacity, which meant the stream was ecologically damaged.

The Politicization of an Environmental Issue

The Suwoncheon covering construction turned into a big political issue in the city at that time. It is important to note that the issue was not one of critical danger in that it was not a life-or-death issue for the local residents. The environmental problem, however, became a political issue and eventually changed an ordinary industrial city into an eco-city. This section points out the critical events, which show how a relatively small environmental issue can be politicized in local society.

First Nationwide Local Elections (1995)

The first simultaneous nationwide local elections were held in 1995 in South Korea and two of the Suwon mayoral candidates had distinctive views on the Suwoncheon covering construction. The then-ruling Democratic Liberal Party's (DLP) candidate, Mr. Lee Hosun, supported the construction plan in order to solve the problem of heavy traffic congestion in the city center. However, the opposition candidate, Mr. Sim Jaeduek, an independent candidate without political party affiliation, was strongly against the construction. Mr. Sim believed that covering the Suwoncheon would not help with the congestion problem but instead would severely damage the cultural and historical monuments in the surrounding area and also impact on the ecological environment of the Suwoncheon (Yoem 2012).

Mr. Sim became the first elected mayor of Suwon and his administration wanted to overturn the plan to cover the Suwoncheon. However, it was not easy to stop the construction because firstly, it was already underway and was already partially complete, and secondly, as mentioned above, the plan was part of a national project pushed by then President Roh to develop and rejuvenate the old city center, and this plan had been one of his presidential campaign promises (Yoem 2012).

Opposing the Local Government

The first official civilian movement opposing the local government came from the Gyeonggi Historical Society, an association of university professors in the Gyeonggi province area. In October 1995 the society issued a joint statement calling for the "suspension of the Manseog Bridge reclamation and covering construction of Namsumun"[1] (Yoem 2012, para 4) due to its historical importance.

1 수원에서는 이후 국책사업을 중지시키기 위한 시민운동이 시작됐습니다. 1995년 10월 경기지역 대학교 사학과 교수 모임인 경기사학회(회장 박성봉, 경희대)가 "만석거 매립과 남수문 복개 계획의 중지"를 요구하는 성명을 내며 불을 지폈습니다.

Local NGOs also took action simultaneously but focused more on the Suwoncheon covering construction. Fifteen local NGOs, including the Suwon Citizens Coalition for Economic Justice (SCCEJ), the Suwon YMCA, and others, got together and formed a joint environmental movement action group, the Suwoncheon Revive Civic Movement Headquarters. This action was led by an individual, Mr. Yoem Taeyoung, who studied and worked in the environmental sector in Suwon city and in 1994 had established the local environmental NGO, Suwon Center for Environmental Movement. The environmental movement action group held a press conference in January 1996 calling for the "stopping of the Suwoncheon covering construction" and "demanding the restoration of Namsumun" (Interviewee 1).

The issue of pollution triggered strong reactions in the local residents and eventually the Suwoncheon Revive Civic Movement Headquarters led a demonstration against the Suwoncheon covering construction. These activities could only be done with strong support from the local residents. Actions such as organizing press conferences, demonstrations, and other relevant activities continued for several months in Suwon city.

Confrontation within the Local Society

Not only were contestation and complaints directed towards the local government. The environmental activist group movement also created confrontation within the Suwon local society. A member of the local NGOs described the Suwoncheon covering construction issue as a hard battle between two distinct interest groups in Suwon city. The group supporting the construction included small business owners in the city center and in the traditional markets, who expected the plan to have direct impact on traffic congestion and parking spaces.

In particular, small business owners from Nammun market were most strongly in favor of the construction due to the fact that Nammun market is the most centrally located in the city and land prices there are the highest in Suwon city (Kim 2002, 48). Parking space was a crucial issue for small business owners in the traditional market because more space would bring more customers, and for this reason they did not want the plan to be revoked (Interviewee 1).

On the other side stood the NGOs and local residents who were concerned about ecological issues in the city. They saw evidence of the pollution problems in and around the Suwoncheon basin and thus wanted the construction to be stopped immediately (Interviewee 3). That is why the NGOs formed the joint action group, the Suwoncheon Revive Civic Movement Headquarters.

External Mediator

As stated, the conflicts were based on disagreements between two distinct interest groups: local business representatives and the environmental action group. Neither side wanted to give in to the other and it seemed there was no solution. A number of meetings and hearing sessions were organized by the Suwon local government and the environmental NGOs, but local business representatives and the environmental action group were unable to reach an agreement.

The environmental action group then decided to send an official request to the Korea Cultural Heritage Bureau (now the Korea Cultural Administration). The action group asked the Bureau "whether Suwoncheon's covering construction would damage the cultural value of the Fortress's Namsumun area"[2] (Yoem 2012, para 6). The Bureau responded by saying that "the Suwoncheon covering construction would degrade the historical and academic value of the Hwaseong Fortress" (Yoem 2012, para 6; Interviewee 3).

This official answer from the Cultural Heritage Bureau changed the dynamic of the conflict between the two interest groups. Due to the fact that Suwon city was already preparing for the Hwaseong Fortress to be listed as a UNESCO World Heritage site, it would have been a deal-breaker if the fortress were to be damaged. A large number of local residents preferred that Hwaseong Fortress be listed as a UNESCO World Heritage site rather than have a parking space in the city center over the Suwoncheon basin area. Suwon inhabitants believed that Suwon city's reputation would increase if it were awarded the title and eventually that would lead to increasing numbers of tourists to the city, which in turn would bring more business to the group of small business owners in the city center (Interviewee 3).

In May 1996, Mayor Sim officially announced cessation of the Suwoncheon covering construction, seven months after the contestation began (Yoem 2012).

Second Nationwide Local Elections (1998)

The Suwoncheon covering was still a hot issue at the time of the second nationwide local elections in June 1998. Mayor Sim again stood as a candidate to be re-elected without political party affiliation, and Mr. Lee stood for a

2 시민운동본부는 문화재관리국에 "수원천 남수문 터 복개 중지 및 원형 복원 요청"을 탄원하고, 문화재관리국으로부터 "수원천 복개는 화성의 역사적, 학술적 가치를 저해한다"는 답변을 이끌어냈습니다.

second time as the candidate of the United Liberal Democrats (ULD) party. During the election campaign, Mr. Lee again argued that the Suwoncheon covering construction was necessary and thus the two candidates engaged in a public debate on the issue. The election of Mayor Sim for a second term meant that the covering of Suwoncheon was stopped for good, which then opened the possibility for a new political agenda in favor of restoring the stream (Yoem 2012).

Policies around the Suwoncheon stream, which pertained to utilization of the area for economic development and the subsequent process of stopping the development, became testimonies for the citizens that environmental decisions should be made with caution and with a long-term viewpoint, and that citizens must all participate in such decision-making processes.

During this period, Suwon residents developed a notion of the importance of the natural environment and how that is connected to the local government, which makes decisions for the local society and also promotes awareness about the role of ecology in everyday life (Interviewee 1; Interviewee 3). This proved that the Suwoncheon covering construction controversy brought new knowledge and educational learning experiences to local residents. Moreover, Suwon residents also understood that by taking action on environmental issues collectively they could change the outcomes of a local election.

If the local residents had not taken action and had not paid attention to this Suwoncheon issue, the result would have been different. The concrete covered part would still be used for parking spaces and the Suwoncheon would have been ecologically damaged, particularly the flora under the concrete covered part that did not have access to sunlight. Instead, the local residents took joint action, and this became the foundation for the transformation to sustainable city development, which is led by the local government in Korea (Interviewee 4).

Announcement on the Demolition of the Suwoncheon Covered Structure

A new mayor, Kim Yongseu, was elected in Suwon city in June 2002, in the third nationwide local elections. His main slogan was "Happy Suwon," which was intended to represent an aspiration to improve the overall quality of Suwon city. One of his important achievements was the extension of and improvements to Suwon's roads. Mayor Kim's local government also built several educational institutions, libraries, and the like in the city. Based on these actions one would assume that as mayor he was more in favor of economic development (Interviewee 1; Interviewee 3).

Despite these developmental projects, Mayor Kim also initiated and implemented several environmental projects in Suwon, particularly with regard to Suwoncheon. One such important decision was to demolish the concrete covering of the Suwoncheon in October 2005. Having stopped the Suwoncheon covering project, this decision represented one of the most cherished ambitions of environmental NGOs and local residents in Suwon city. Indeed, although local residents had originally welcomed the plan to cover the stream, as mentioned above, they soon realized the negative environmental impact. It was for this reason that the first mayor, Mr. Sim, was re-elected through a direct voting system in 1998, and the new mayor, Mr. Kim, could not completely ignore this fact.

Some locals suspected that he made a number of important environmental decisions in Suwon city due to the appearance of a political rival, Mr. Yoem Taeyoung, who was the former head of a local environmental NGO and led the Suwoncheon restoration movement successfully in the past (Interviewee 2). Although this information was not confirmed by Mr. Kim himself, it should be noted that some important environment-related decisions were made just several months before the fourth and fifth simultaneous nationwide local elections. For example, in July 2006, Suwon agreed to a sister city relationship with the Brazilian city of Curitiba, which is well-known throughout the world as an eco-city (Suwonshi 2015a, 109). Furthermore, it should be noted that demolition of the Suwoncheon covering was announced roughly seven months before the fourth nationwide local elections, which were planned in June 2006.

Suwon as an Eco-City

The making of Suwon as an eco-city, from an industrial city, was a major transformation. The process required a combination of strong leadership, efficient work by the public administration, and support from and active involvement by local residents. In particular, local government leadership meant that the elected mayor's role was one of the crucial elements in the transformation to a sustainable city. In the case of Suwon, interviews also indicated that environmental awareness among local society has been an important element of the changing local politics. Moreover, good cooperation among several institutions as well as with residents has brought together local government, quasi-government institutions, and NGOs to transform the city.

It is important to note, however, that this local transformation occurred in the context of national policies (and indeed also of emerging international frameworks). Initiatives from the central government and successive presidents provided opportunities for local leaders to engage in the kind of policies that would ultimately transform the city of Suwon. The argument made here is therefore not that local leaders were single-handedly responsible for the outcome. Rather, the research presented here demonstrates that such transformation relied on local responses to national policies, and that local leadership was essential in bringing about the change that has been observed. In other words, while not providing an exhaustive account of the reasons for transformation, this subsection answers the questions of "why here?" and "why now?"

Electoral Politics as a Background to Suwon Eco-City

In the fifth nationwide local elections of June 2010, Mr. Yoem Taeyoung stood as a candidate for mayor. For this election, he was nominated by the opposition party, and won against a candidate from the ruling presidential party. Mr. Yoem had been the leader of a local environmental NGO and led the environmental movement against the Suwoncheon covering construction in the mid-1990s.

In these local elections, Mr. Yoem's campaign made "Five Pledges to Suwon" voters, including the promise to work actively for "economic balance" in Suwon. He outlined an environmental plan that included the Hwaseong Fortress restoration project, environmental improvement of residential areas, passing new ordinances to protect small businesses, and creating a shopping district in the city center. However, in his pledge, Mr. Yoem did not use the terms environment or ecological protection directly—restoring the historical monument and developing the city center would be accompanied by environmentally friendly action based on his past professional background (Yoem 2010).

However, it has been argued that Mr. Yoem built his reputation around the environmental issue in Suwon and projected a positive impression and image as a successful environmental activist based on his involvement in the Suwoncheon issue in the mid-1990s. During in-depth elite interviews, it was claimed that voting for Mr. Yoem meant that Suwon residents sought improved environmental and ecological living conditions in the city. As such, it was not necessary to specify the environment or protection in his election manifesto. Local residents had assumed that Mr. Yoem would prioritize the environment in the municipal government's affairs (Interviewee 1; Interviewee 3). Mr. Yoem won the election and became the third mayor of Suwon at the fifth nationwide local elections held in June 2010.

The Decision-Making Process of the Eco-City-to-Be Suwon

So far, the two most important elements identified in the transformation of Suwon to an eco-city were the politicization of the environmental issue and the (high) profile of the elected politician. In fact, those two elements are connected somewhat in a cause-effect relationship. Initially, Suwoncheon's environmental problem was not a critical issue that would affect the local residents' daily lives. But politicization of the environmental issue was successful, resulting in it being integrated into the local political arena from the emergence of the Suwoncheon issue, which eventually influenced the outcome of the local election.

The decision-making process in the transformation of Suwon to an eco-city happened rather quietly and quickly, without any official announcement or visible external event. However, the term "Environmental Capital" appeared almost immediately after the electoral result was published, when the title was used by the Transition Committee.

Thus, it can be argued that the concept of establishing or transforming Suwon into an eco-city was already intended and outlined, even before the official local election campaign. A number of interviewees confirmed this contention. They responded that Mr. Yoem was known in the local society as an environmental activist. By giving him the vote, this meant there was mutual understanding in support for sustainable city planning, which contributed to smoothly establishing plans for the eco-city. Thus, when the opportunity presented itself after the local elections, the expectation was that the project would proceed without hesitation.

Transition Committee

Almost immediately after the election result was announced, a Transition Committee was formed to ensure the smooth handover of power from the previous mayor (Interviewee 7). Mr. Yoem nominated the members of the Transition Committee, who were grouped into five different Task Force teams with different focal points linked to the five election pledges. The five Task Force teams were focused on (1) employment, (2) ecologically friendly free school meals, (3) city center regeneration, (4) Environmental Capital, and (5) citizen participation.

This was the first time the term "Environmental Capital" appeared officially in Suwon city. One of the Transition Committee members revealed that the new mayor and relevant actors wanted Suwon city to become an important and central city in Korea with regard to environmental concerns. Although the term *Hwangyeongsudo* (Environmental Capital) had not been certified by any organizations or individuals, and Suwon city had not had the intention to do so, it was hoped that Suwon city would become a center of environmental sustainability in Korea (Interviewee 1; Interviewee 2).

The Transition Committee was dissolved about a month after Mayor Yoem's inauguration. However, the Task Force teams were transformed into another organization within the Suwon local government, namely the Suwon Municipal Advisory Council (Interviewee 5).

Green City Suwon

On the issue of the environment, the NHS Trust Report stated the objective of "a vibrant environmental city" and proposed three strategies: "to improve the urban environment," "to construct a green transportation system," and "to realize a low-carbon green city"[3] (NHS Trust 2010). Based on these objectives, strategies, and promises, the Transition Committee outlined an "Implementation Roadmap"[4] and prepared a budget (NHS Trust 2010, 44). The Implementation Roadmap specified the timeline of strategies during the mayor's term from 2010 to 2014. It also included which department(s) and individual(s) would take responsibility for the implementation process and which departments and individuals would be indirectly involved with the process.

Establishing the new image for the city was not straightforward. The catchphrase of Suwon as an ecological city changed over time. For Suwon to become the "Environmental Capital," decisions had to involve several actors, namely bureaucrats from the Suwon government, members of former and current environmental NGOs who worked together with Mr. Yoem, and local business representatives in and around Suwon city.

Declaration of Environmental Capital Suwon

Together with other representatives from various organizations in local society, Mayor Yoem declared the "*Hwangyeongsudo* (Environmental Capital) Eco-City Suwon" in September 2011. The membership of the declaration committee indicated that the Eco-City Suwon project was not only the

3 "생동감 넘치는 환경도시," "도시환경 개선," "녹색교통체계 구축," "저탄소 녹색도시 구현."

4 "로드맵 구현."

work of the local government but also involved various actors from Suwon's local communities.

In the Environmental Capital Suwon declaration, the promise was made to reduce greenhouse gas (GHG) emissions by 40 percent by 2030 compared to 2005. While using the symbolic word *Hwangyeongsudo* in their slogan, Suwon clarified its aim of becoming a Low-Carbon Green City (NHS Trust 2010). This title indicated more clearly the goals of Suwon city as it also scaled down the implementation plan and provided a more realistic and achievable plan.

While the decision to designate Suwon as an eco-city originated in local politics, a strong international dimension emerged, as it also occurred in the context of a national drive to pay greater attention to environmental objectives. In 2008, then President Lee declared the Five-Year Plan for Green Growth National Strategy (2009–2013). This national strategy aimed to provide a "low-carbon, green growth model" to direct the nation to achieve long-term economic development (Kamal-Chaoui et al. 2011, 8).

The Administrative Process towards an Eco-City

Changing the concept of the city was a long-term process requiring time, resources, and the capacity of the municipal government. Suwon's local government also made internal and external preparations that included legal action. One of the most important decisions was to restrict the Suwon government's organizational, including adding a second vice-mayor. Thus the organizational chart was rearranged around two vice mayors and the Environment Division and City Planning Division were assigned to Second Vice-Mayor Lee Jaejoon. Those two divisions were highly important departments for the Environmental Capital project and therefore the second vice-mayor became one of the crucial figures in its planning, implementation, and execution (Interviewee 6).

The second change would be in establishing new departments to take responsibility for areas that are highly relevant environmentally, such as the water and sewage systems, waste management, and energy generation. New departments were instituted and assigned tasks such as "organizing environmental events," "aiming for the environmental international award," "building the energy usage monitoring system," and other relevant tasks linked to the Environmental Capital project[5] (Gugtoyeonguwon 2012, 261).

5 "환경 관련 각종 국제행사 개최," "국제 환경수도 어워즈(가칭) 추진," "에너지 사용량 모니터링 시스템 구축."

Legal Preparations

The Suwon municipal government established and revised the relevant legal procedures so that the Environmental Capital project was achieved within the existing legal boundaries. The local government of a large city which serves more than 500,000 residents can set its own legal framework on four different levels, namely, in order of significance, *Jolye* (ordinance), *Gyuchig* (regulation), *Hunlyeong* (instruction), and *Yegyu* (established rule).

In November 2016, Suwon municipal government had 429 Ordinances, 136 Regulations, 60 Instructions, and 19 Established Rules (ELIS n.d.). This legal framework was spread over 17 sectors from the City Council, Environment Divisions, and local administration divisions, *Gu* and *Dong* (ELIS n.d.). In order to proceed with the Environmental Capital (Low-Carbon Green City) project, the Suwon government instituted new ordinances and also revised existing ones. Most of the ordinances that related to the project pertained to the environment.

International and Domestic Benchmarking

In order to establish the Environmental Capital (Low-Carbon Green City) designation, the administration had to consider the international dimensions of such a process. The first approach was to study other developed countries such as the United Kingdom, Japan, and Germany to find out how these countries tackle the problem of GHG emissions and establish future targets. The second was to search other examples of environmental capitals in the world. The administration's study showed that there were several environmental capital schemes, which were organized by different institutions. The third approach involved research on individual cities with good reputations with regard to sustainable and environmental development, such as Ulm in Germany, Viikki in Finland, Hammarby Sjöstad in Sweden, Masdar in the United Arab Emirates, and Dongtan city in China (Gugtoyeonguwon 2012, 234).

The domestic benchmarking involved both central government and local government initiatives. The Saengsaengdoshi (Eco Rich) City program was one of the central government initiatives. Suwon carefully assessed the evaluation criteria and adapted these into the Environmental Capital (Low-Carbon Green City) project, and as a result was awarded the Presidential Prize of Eco Rich City in 2011. The second central government initiative was the Climate Change Adaptation Model City project. The Ministry of Environment (MoE) supervised this project from 2007 to 2009. The project specifies the GHG emissions reduction target for each local government.

Making an Eco-City Master Plan

While the Suwon local government was preparing for the Environmental Capital project, it realized that it was necessary to get external professional input. The project required additional work by local officials on top of their daily tasks and they needed experts' knowledge to realize the plan properly.

Research of the local government's official documentation archive revealed two different aspects of external involvement in the Suwon city Environmental Capital project. The members of Green Korea and Green Energy Korea Cooperation prepared a document titled "The Environmental Capital (Low-Carbon Green City) Master Plan – Establishment of Citizens' Participatory Governance," which was submitted to Suwon city in February 2012. In particular, the NGOs organized several information sessions, discussion sessions, and meetings to invite local residents to inform them about the project and discuss the aims and action plans in advance. This would be one of the positive points of the Suwon Environmental Capital project. Without the support and cooperation of the local residents, it would not be possible to achieve the long-term environmental project.

After that, Suwon city outsourced research for the Environmental Capital Master Plan to a think-tank called the Korea Research Institute for Human Settlements (KRIHS). KRIHS prepared the report "A Comprehensive Plan for Environmental Capital – Low-Carbon Green City," which featured recommendations in designing a long-term vision, clear goals, tactical strategies, and achievable milestones for 2010 to 2014 and beyond. Moreover, a number of local NGOs also participated in the preparation of that part of the report dealing with local governance (Gugtoyeonguwon 2012).

Eventually, the Master Plan was completed in February 2012. After the completion of the Master Plan, the actual project, including building construction began in March 2012 (Suwonshi 2016). The Suwon Environmental Capital Master Plan outlined three objectives: "Ecologically friendly environmental city," "A safe low-carbon green city for climate change," and "Citizens participating in city governance"[6] (Gugtoyeonguwon 2012, 178). Each of the objectives included three different strategies, and the strategies in turn contained three to four different action plans so that project implementation would be achieved step by step in each category.

According to the Suwon Municipal Administration White Paper of 2015, eight elements of the action plans were completed from 2011 to 2014 and the

6 "자연 친화적인 환경도시 추진," "기후변화에 안전한 저탄소 녹색도시 조성," "시민이 참여하는 거버넌스 도시."

programs were managed by the relevant institutions and actors in Suwon (Suwonshi 2015a). Those completed action plans were "the ecological water stream restoration project," "establishment of the 2030 Low-Carbon Green City Master Plan," "establishing a long-term and future-oriented plan for West Suwon," "establishing the Suwon Atopic Skin Disease Center," and "establishing the Suwon City Eco Center."

Other important achievements included the "reduction of GHG process" and the "Eco-Mobility World Festival 2013," which was co-hosted by Suwon city, ICLEI Local Governments for Sustainability (ICLEI), and UN-HABITAT (Suwonshi and ICLEI 2014).

Conclusion

This chapter traced the process of how an ordinary industrial city, Suwon, transformed itself into an eco-city. Suwon city was declared "*Hwangyeongsudo* (Environmental Capital) Eco-City Suwon" in September 2011. It is important to note that, while the trigger for this development lay in the contestation over the stream, the implications—economic, ecological, and political—have been much broader. In his declaration, Mayor Yoem stated that the aim was to reduce greenhouse gas (GHG) emissions by up to 40 percent by 2030 compared to the levels of 2005, demonstrating the wide-ranging ambitions of this initiative. The research traced why and how this event could take place and what happened afterwards in order to fulfill the declaration.

The research demonstrates that the transformation process had its origins in the mid-1990s with the environmental conflict over the Suwoncheon. Predominantly, the conflict was about the covering of the Suwoncheon water stream and the negative impact that this construction had on the local environment and on the city's cultural heritage.

The debate over the covering of Suwoncheon with concrete, the removal of these concrete covers, and the subsequent restoration of the basin area took nearly two decades and was eventually completed in 2012. During this time, the Suwoncheon issue provided an opportunity for the local population to appreciate the importance of the ecological environment in the context of urban living. This learning experience also triggered a change in the local society, and, as a result, facilitated a chance for locals to elect an environmental activist as their mayor, an individual who had previously led the environmentalist movement in Suwon.

The research identified a number of factors that were already present before the beginning of the decision-making process that led to Suwon becoming an eco-city. In the case of Suwon city, the two most important factors were identified as the politicization of the environmental issue and the electoral politics of the city. Importantly, those two elements were connected to one another because Mr. Yoem gained popularity and was already known by the local residents due to the Suwoncheon conflict. The Suwoncheon issue was well politicized and eventually provided the springboard for an environmental activist to become the city's mayor.

The decision-making leading to the transformation of Suwon to an eco-city happened comparatively swiftly, without the long and time-consuming process that would normally be expected. The term "Environmental Capital" was introduced when Mr. Yoem appointed the members of the Transition Committee, and following his inauguration, the eco-city transformation project immediately went into action. The idea to be recognized as an eco-city was put forward by the members of this Transition Committee. The Suwon municipal government also supported this process and worked together with the committee from the outset, a process that then continued permanently.

In order to transform an ordinary industrial city into an eco-city, the relatively quick decision-making to proceed with the project was followed by a sustained process of implementation. This chapter has traced a number of changes in the municipal government that occurred in this context, such as, the creation of the new position of second vice-mayor and reforming the organizational structure so that the Suwon government could work effectively on the Environmental Capital project. A number of new legal elements were also introduced by the municipal government in order to proceed with the project. The new ordinances focused on low-carbon green growth, a number of environmental education promotions, and the establishment of an environmental atopic skin disease center. Furthermore, external professional input was required due to both the additional work required from local officials in order to manage the new project on top of their daily tasks, and the need for expertise to realize the new plans.

It is also important to point out what has changed since the implementation of the Environmental Capital project in 2010. The number of changes is observable—the most obvious result is the completion of the Suwoncheon restoration project, which took over a decade. Now the Suwoncheon basin area has been turned into a small park with a scenic zone where Suwon residents can enjoy a walk and go jogging. The Suwon municipal government completed eight different action plans including the completion of the Suwon Atopic Skin Disease Center and the Suwon City Eco Center, both of which are evidence of the new physical environment in Suwon city.

Most importantly, the air quality in the city has improved—as the Suwon mayor promised—as GHG emissions levels were reduced to 3.2 percent in 2014, compared to the 2005 level. Suwon claimed that GHG emissions were reduced by 860,000 tons in 2015 (Suwonshi 2016, 682). In short, Suwon city declared the partial completion of the Environmental Capital project in 2014, though further actions will be required to achieve the full completion of the eco-city project by 2030.

The Suwon case has revealed that it takes time to transform an ordinary city into an eco-city in terms of developing the plans, proceeding with the implementation process, and completing the project. It also involved the mobilization of the local population, the creation of a dedicated environmental movement, and the involvement of civil society in local politics—all developments that not only changed the outcome in terms of a new balance between ecological concerns and economic interests, but also created new and lasting dynamics in terms of greater citizen involvement in urban governments.

While these are potentially profound changes, the actual decision-making process in this particular case was comparatively uncomplicated. With political power being concentrated in the hands of local mayors in South Korea, they are able to introduce new initiatives and change a city's future plan. Therefore, local residents have the power to express their preferences at the local election and in that way shape and—when and where necessary—change the course of local development in recognition of new preferences such as, in this case, the protection of the natural habitat and cultural heritage related to the city's river.

Acknowledgments

The author would like to thank all interviewees who participated. In total, 14 interviews were conducted with Suwon local actors who were directly involved with the eco-city transformation process. All interviewees remain anonymous, in line with the commitments made in the course of the research.

References

Bennett, Andrew, and Jeffrey Checkel (eds.). 2014. *Process Tracing from Metaphor to Analytics Tool*. Cambridge: Cambridge University Press.

Collier, David. 2011. "Understanding Process Tracing." *PS: Political Science and Politics*, 44(4): 823–830.

Dexter, Lewis. 1970/2006. *Elite and Specialized Interviewing*. Evanston, IL: Northwestern University Press (1970); Colchester: ECPR Press (2006).

ELIS. n.d. "Gyeonggi Province – Suwon City." *Enhanced Law Regulation Information System* (ELIS), Ministry of Interior of South Korea. Available at http://elis.go.kr/newlaib/laibLaws/h1126/laws.jsp?regionId=41110&isfez=, accessed 5 November 2016.

Gugtoyeonguwon. 2012. *Hwangyeongsudo (Jeotanso Nogsaegdosi) Gibongyehoeg – Choejongbogoseo* [Environmental Capital (Low-Carbon Green City Master Plan)]. Research Commissioned by the Suwon Municipal Government, Gugtoyeonguwon (Korea Research Institute for Human Settlements, KRIHS). Anyang: KRIHS.

Hu, Mei-Chih, Jessica Wadin, Lo Lagerstedt, Chen Hsien, and Jian Yuan Huang. 2016. "Transformation toward an Eco-City: Lessons from Three Asian Cities." *Journal of Cleaner Production*, 123: 77–87.

Johnson, Björn. 2007. "System of Innovation, the Urban Order and Sustainable Development." *Waste Management & Research*, 25: 208–213.

Kamal-Chaoui, Lamia, Fabio Grazi, Jongwan Joo, and Marissa Plouin. 2011. *The Implementation of the Korean Green Growth Strategy in Urban Areas, 2011/20*. OECD Regional Development Working Papers. Paris: OECD Publishing.

Kenworthy, Jeffrey R. 2006. "The Eco-City: Ten Key Transport and Planning Dimensions for Sustainable City Development." *Environment and Urbanization*, 18(1): 67–87.

Kim, Hyeonmi. 2002. "Suwoncheoneun Sala Heulleoyahanda [Suwoncheon Should Flow]." *Jugan Donga Weekly Journal*, 12 December. Available at http://weekly.donga.com/List/3/all/11/69421/1, accessed 3 November 2016.

Kwon, Sangeun. 2008. "Suwoncheoneul Cheonggyecheoncheoleom Mandeunda [Make Suwoncheon Like Cheonggyecheon]." *Joseonilbo Newspaper*, 19 February 2008. Available at http://www.chosun.com/site/data/html_dir/2008/02/19/2008021900029.html, accessed 23 November 2016.

Lee, Bum Hyun. 2012. *2011 Modularization of Korea's Development Experience: Korean Version of New Town Development*. Ministry of Land, Transport and Maritime Affairs (MLTM), Republic of Korea. Available at http://ksp.kdi.re.kr/publication/modul.jsp?syear=&sage=&skey=&stem=&stype=&pg=5&idx=57, accessed 3 May 2016.

Lee, Jong Youl, and Chad David Anderson. 2013. "The Restored Cheonggyecheon and the Quality of Life in Seoul." *Journal of Urban Technology*, 20(4): 3–22.

Leznicki, Marcin, and Aleksandra Lewandowska. 2014. "Implementation of Sustainable Development on the Example of the Concept of Eco-City." *Ecological Questions*, 19: 91–96.

Lye, Liang Fook, and Gang Chen (eds.). 2011. *Towards a Liveable and Sustainable Urban Environment: Eco-Cities in East Asia*. Singapore: World Scientific Publishing Co.

NHS Trust (Name Withheld). 2010. *Hyumeonsiti Suwon, Salami Bangaun Dosi.* Unpublished confidential document at the Suwon Municipal Government, Suwon.

Roseland, Mark. 1997. "Dimension of the Eco-City." *Cities*, 14(4): 197–202.

Suwonshi. 2015a. 2015 *Suwon Sijeong Baegseo* [Suwon Municipal Administration White Paper 2015). Suwon City: Municipal Government Printer.

Suwonshi. 2015b. 2014 *Suwonshi Hwangyeongbaegseo* [Suwon Municipal Administration Environmental White Paper 2014]. Suwon City: Municipal Government Printer.

Suwonshi. 2016. *Suwonsi Hwangyeongbojeongyehoeg 2016–2025* [Suwon City Environmental Conservation Plan 2016–2025]. Research Commissioned by the Suwon Municipal Government from Aju University. Suwon City: Municipal Government Printer.

Suwonshi and ICLEI. 2014. *EcoMobility World Festival Suwon 2013: A Pleasant Walk in the City.* Suwon City Transportation Policy Division (Eco Mobility Team) and ICLEI-Local Governments for Sustainability Korea Office. Suwon City: Municipal Government Printer.

Tight, Malcolm. 2017. *Understanding Case Study Research: Small-Scale Research with Meaning.* London: Sage.

UNESCO. 1997. *Hwaseong Fortress.* United Nations Education, Sciences and Cultural Organization, World Heritage Convention. Available at http://whc.unesco.org/en/list/817/, last accessed on 3 November 2016.

Wong, Tai-Chee, and Belinda Yuen (eds.). 2011. *Eco-City Planning: Policies, Practice and Design.* New York, NY: Springer.

Yoem, Taeyoung. 2010. *Huboja Haegsimgongyag. Seongeogongyagseo Jechulmich Gonggae Seungnagseo* [Candidate Key Election Pledges and Public Standing to Consent Form], submitted to Suwonsipaldalguseongeogwanliwiwonhoe (Suwon City Paldal District Election Commission) on 15 May 2010.

Yoem, Taeyoung. 2012. "2012 Sijeongilgi: Suwoncheon bogwon Siminundongi Ilun Nollaun Yeogsaibnida [2012 Municipal Administration Diary—Suwoncheon Restoration Is an Amazing Historic Accomplishment of the Civil Movement)." Suwon City Mayor Mr. Yoem's official blog. Available at http://blog.daum.net/lovesuwon/1023, last viewed on 1 November 2016.

Biography

Dr. Youngah Guahk is a political scientist working as a lecturer of Asian affairs at LUISS University in Rome, Italy.

9 Water's Edge Urbanisms along the Buckingham Canal in Chennai

Karen Coelho

Abstract

This chapter examines the edges of urban waterways as intra-urban peripheries where dynamic processes of place-making and appropriation occur. Chennai's Buckingham Canal has undergone shifting valuations in the city's urbanization history since its construction by the British as a navigation channel in the late 1800s. Having degenerated into an urban drain after the 1950s, it was revalorized in the 2000s through new eco-restoration visions. This dynamic ignored other histories of the canal including those of informal settlements and small-scale industries that had grown along its banks, offering working-class families a foothold in the city, and who now face eviction. This chapter draws on multi-sited ethnography in three canal-bank settlements to describe the shifting trajectories of opportunity, challenge, and threat that water's edge urbanism offers.

Keywords: Chennai, water's edge urbanism, Buckingham Canal, urban political ecology, periphery

Introduction: Water's Edge Urbanisms

Drawing on a case study of Chennai's Buckingham Canal,[1] I propose in this chapter that the banks and shores of urban waterways constitute a

1 This chapter is based on a case study of settlement along the Buckingham Canal (Coelho et al. 2017), commissioned by the Indian Institute of Human Settlements (IIHS) in Bangalore. The full case study is available at http://cases.iihs.co.in/#casearchive/-KW1dn-bB93VvMVD2e2r. My thanks are due to Gautam Bhan of IIHS, and to the team of researchers in Chennai that conducted and wrote up the ethnographies for this study: T. Venkat, Meghna Sukumar, Anusha Hariharan, Madhura Balasubramaniam, M. Subadevan, and R. Vaishnavi.

Padawangi, Rita, Paul Rabé and Adrian Perkasa, *River Cities in Asia*. Taylor & Francis Group, 2022
DOI: 10.5117/9789463721851_CH09

type of intra-urban periphery, giving rise to distinctive forms of subaltern urbanism that I term "water's edge urbanisms." This argument rests on the unregulated tenurial status of water's edge spaces in Indian urban centers. These spaces, constituted by non-revenue lands known in southern India as *poramboke*,[2] are typically subjected to a range of informal usufructs. The term subaltern is used here to propose that groups marginalized by dominant engines of urbanization (fueled by the logics of real estate and finance), or who have no access to formal or legal rights to the city, build urban spaces and cultures that counter—often without explicit intent—such hegemonic urban projects. They typically do so in defensive, subversive, and often undercover modes.[3] Water's edges, then, are spaces where dynamic processes of urbanization can be traced over time. Informal or para-legal settlement leads to the spontaneous and incremental production of highly diverse zones of urban occupancy—residential, industrial, commercial, or mixed—followed almost inevitably by contestations and struggles over various forms of appropriation or threat, including eviction, polluted/insanitary conditions, and flooding.

A growing body of writings has begun to examine the ways that flows and stores of water (in rivers, tanks, wells) configure social lives, generate value, and provide the foundational features for the establishment, design, and dynamics of urban settlements (Hastrup 2013; Shannon 2013; Mathur and da Cunha 2009). These writings, which I categorize here under what Shannon (2013) calls "water urbanism," focus on water as a uniquely agentive element or substance, and explore how its forms, functions, movements, and meanings shape urban social formations. Hastrup (2013), for instance, argues for a closer engagement with "water cultures," with how water is socially experienced in its fluidity and circulation. Mathur and da Cunha (2009) interrogate the contested boundaries and negotiated relations between cities and water, and advocate for approaches that accommodate rather than control the ambiguity and complexity of water.

2 The term *poramboke* designates a category of non-cultivable land that had been customarily reserved for public or communal purposes (including grazing lands, forests, and floodplains) under the authority of the village administration. In current times this category is broadly glossed as "government-controlled commons" and in the case of water's edge *poramboke*, is controlled by the Public Works Department.

3 This usage of the term subaltern draws on Gramsci and postcolonial studies, to signify populations that are positioned outside hegemonic power structures or officially sanctioned categories. Denied agency by this positioning, they occupy "spaces of difference" or engage in counterhegemonic practices.

The framework of water's edge urbanisms that I propose here builds on this rubric of water urbanism, but differs from it in two ways. First, my concern here is with subaltern urban settlements on the banks of waterbodies, thus not as much with water as with land, however contingently defined and bounded. And it deals with a specific category of socially produced urban spaces that are marked and shaped—ecologically, economically, culturally—by water. These are spaces whose transformations and changing values are determined by their proximity to and relations with water.

Second, in contrast to the design-oriented approach of water urbanism,[4] the plural concept of "urbanisms" here denotes the varied ways (some of which will be elaborated later in this chapter) in which urban spaces on water's edges are historically produced by subaltern urban groups as distinctive places. Inflected by their boundaries with water, these places are subjected to changing valuations in dominant metrics of urban development. Place-making[5] here unfolds within a dialectics of value, in which these spaces figure for long periods as backyards, invisible to capital and sidelined by the state, until they are propelled into prominence as frontage or as high-value real estate.[6] Water's edge spaces thus emerge as a periphery, an "elsewhere" to dominant processes of city-making, even while such processes contribute to their eventual appropriation through various forms of state action.[7] Importantly, these are intra-urban peripheries. In this sense they are more than the fringes of the urban mainstream or—as scholars have characterized urban peripheries—a seam connecting "natural" and built space (Goldman and Narayan 2019), the authorized and unauthorized city, or the accomplished city and the city-yet-to-come (Simone 2004). The water's edge as intra-urban periphery is also a corridor linking diverse settlement ecologies and various kinds of localized urbanisms within a single city.

4 Shannon (2013, 163) describes water urbanism as "the design of cities around the role and dynamics of water." This paradigm of "designing with water," under pressure to adapt to changing social and climate imperatives, is, in her view, evolving from hard to soft engineering practices.

5 The concept of place-making here signifies the everyday practices that materialize space, drawing from Lefebvre's dialectical treatment of everyday life and the social production of space (see McDuie-Ra 2012). This approach resonates with John Friedmann's (2007) emphasis on the ways that place is made from the bottom-up, in the daily patterns and rituals of life on the ground.

6 For examples, see Coelho and Raman (2010) and (2013) on the changing values assigned to water in the history of Chennai's development, and Baviskar (2011) on changing values of the River Yamuna in Delhi.

7 See Follmann (2016, and this volume) and Baviskar (2011) on how the Yamuna was appropriated by the Delhi Development Authority for various "world-class" construction and real estate projects.

Caldeira's analytic framework of "peripheral urbanization" (2016) offers strong resonances with this concept. For Caldeira, peripheral urbanization refers to a mode of production of space with three distinct characteristics. First, the spaces are auto-constructed, i.e., built from the ground up by their occupants, as opposed to being simply consumed as ready-made commodities (also see Simone 2017). Second, they engage with official logics of capital, property, law, and economy, in a transversal rather than direct fashion since they are typically excluded from formal circuits. And third, they generate new kinds of politics and new citizenship claims. Peripheral urbanization, as Caldeira shows, produces highly heterogeneous spaces that are marked by inequality and precarity. It is also dynamic in that these spaces are subject to constant, if sometimes slow-paced, change. This dynamism occurs both in the form of incremental enhancements in the quality of built space, and of almost inevitable displacement, as the improved spaces are taken over by external actors, state or market, followed by a reproduction of the peripheral urbanization process elsewhere.

The water's edge as periphery adds the crucial dimension of ecology to this framework. Values of people and places here are configured through the prism of "ecological value," a concept which, as recent writings have shown, takes on particular definitions and meanings in different contexts. In contemporary Indian urbanism, ecological values are constituted by greening, beautification and the manufacture of "usable nature" for recreation and aesthetics (Baviskar 2011). In Indian cities since the 1990s, eco-restoration of waterbodies has been inserted into "project" modes, powerfully shaped by considerations of infrastructure financing, waterfront development, and real estate urbanism.[8] As urban waterways—rivers, canals, creeks—become central to projects of world-classing, old settlements on their banks get drawn into shifting dynamics of spatial valorization, from "development" to displacement, which in turn shape their strategies of place-making.

In thus framing the problematic of water's edge urbanisms, this chapter draws on the theoretical tradition of urban political ecology (UPE), which not only posits the inseparability of urban natural, technical, and social formations (see also Follmann, this volume), but allows for an analysis

8 A telling instance of this is the institutional architecture of river restoration in Chennai—all projects are implemented by the Chennai Rivers Restoration Trust, a government-promoted subsidiary of the Tamil Nadu Urban Infrastructure Financing Services Limited, which is a public-private partnership involving private multinational banking and financial services corporations.

of how flows and circulations of capital, nature, discourses, and social processes produce differential values of spaces and people within the city (Heynen 2014; Coelho 2018). As Heynen claims, "UPE research has worked to show in stark terms what is politically at stake via the uneven, and often crippling, socio-natural power relations at play through the urbanization of nature" (2014, 600).

A distinguishing feature of waterways eco-restoration projects in Indian cities, from Delhi to Ahmedabad to Chennai—and indeed across cities of the Global South—is that they selectively target urban poor settlements on the banks as encroachers and as polluters (Mathur 2012; Baviskar 2011; Coelho and Raman 2010, 2013; Coelho et al. 2017; Padawangi 2019; Rademacher 2008). These projects are in fact motivated by, and designed around, the eviction of the urban poor from "prime" urban lands. This is part of their appeal for sections of the urban middle classes who have become increasingly hostile toward the presence of "slums" in the city, seeing them as aesthetic disturbances or eyesores (Ghertner 2015; Coelho and Raman 2010; Baviskar 2002). Eco-restoration projects thus emerge as sites of enormous human tragedy, among the most exclusionary interventions in the contemporary landscape of urban transformation (Coelho 2020).

The already sparse social science literature on urban waterways remains preoccupied with rivers and lakes; canals barely figure in these writings, and where they do (e.g., Hastrup 2013; Mukerji 2009; Wade 1985), they are predominantly presented as grandiose projects of modernist engineering. But canals as waterways are often uncertain, ambiguous figures, on the one hand engineered artifacts expected to deliver economic benefits, and on the other, water flows caught in ecological webs of interaction. Additionally, as infrastructures designed and built at specific historical junctures for specific purposes such as irrigation, navigation, or drainage, they often become awkwardly poised in the rapidly shifting course of urban history, where their values—techno-functional, ecological, aesthetic, or social—may fluctuate or decline.

Chennai's Buckingham Canal offers a productive geography to demonstrate the peripheral dynamics of urban waterways and their borderlands in Indian metropolitan cities. Like other waterways in Chennai, the canal has undergone shifting valuations in the city's urbanization schemes over its history (Coelho 2017; Coelho et al. 2017). Built in the 1800s as a navigation channel to facilitate cargo transportation down the eastern coastal tracts of the Madras Presidency, its operations remained perennially under challenge from storms, tidal action, siltation, and pollution (Russell 1898). By the 1950s, navigation had substantially declined, and the canal had degenerated into a choked and dysfunctional urban drain (Krishnamurthy and Kabalamurthy 1964; GO 1714, PWD, 6/4/1956; GO 3289, PWD, 6/12/1962). Until the 1970s, its banks were seen as a rear in the city's development, a backyard mess. In the 1980s the canal alignment was identified as a tract of cheaply available open space for construction of the city's elevated rail corridor, the Mass Rapid Transit System (MRTS) (MTP 1996).

And in the current millennium, the canal is reimagined as part of a vast inter-state waterway with restored flows and greened waterfront parks, along with a growing discourse about its drainage functions for the city (IWAI 2010; Natarajan 2018; Venkateshwarlu 2017).

These shifting perspectives on the canal have shaped the histories of place-making in neighborhoods along its banks. Settlements along the canal have traveled varied historical pathways to the present. Today, some are concentrated hubs of small-scale manufacturing, others are intensely mixed-use zones of informal commerce, services, and residence. Some have entrenched themselves, been upgraded by the state, and morphed into developed neighborhoods. Others were evicted and resettled elsewhere, still others have remained informal, unserviced, and constantly vulnerable to threats of eviction. Many have served as staging grounds for migrations, voluntary or forced, to other areas of the city. In all these ways, canal-bank settlements have played important roles in shaping the social, cultural, and economic geography of Madras/Chennai.[9]

This chapter explores these processes of mutual production between the canal and the city. The rubric of "river cities" that frames this volume captures this mutual co-constitution of a socio-natural hybrid, an urban system that encompasses waterways and the cities or towns with which they travel. Here, I draw on the UPE approach to explore how historical processes of urbanization in Madras/Chennai have determined the current condition of the canal, and, equally, how the canal's flows and banks shaped the unfolding of particular forms of urban settlement and community that define Chennai's contemporary urban character.

9 In December 2020, as the city found a brief respite from the restrictions of the COVID-19 pandemic and its associated lockdowns, a new project to restore the Buckingham Canal and other drains in the city was announced, which proposed to evict over 17,500 families residing on the banks of these waterways. See https://www.thehindu.com/news/cities/chennai/drive-to-restore-buckingham-canal-scheduled-to-begin-soon/article33375629.ece.

Yet, as this chapter shows, shifts in the valorization of land and water in the city have rendered these settlements vulnerable to demolition for eco-restoration projects. This chapter proceeds by outlining, in the section below, the failure of the canal in its intended purposes of facilitating transport and yielding revenue. These functional metrics have obscured the significant role that the canal has played in the city's social geography through forming the grounds for a range of social and economic trajectories which have provided urban footholds, homes, livelihoods, and socioeconomic mobility to at least two generations of the city's working-class communities. The third section sketches the unfolding of these processes in three distinct settlements on the banks of the canal, highlighting the specific conditions of water's edge urbanisms and the shifting trajectories of opportunity, challenge, and threat they offer in each case. It shows how the production of urban space in each of these neighborhoods was shaped by the canal's changing ecological and economic values over time.

The chapter draws its empirical material from a case study (Coelho et al. 2017) which employed multi-sited ethnography in three settlements along the Buckingham Canal, one each in the northern, central, and southern parts of the city (Figure 9.1). Three research teams spent about three months in each of these settlements, gathering oral histories and conducting in-depth interviews, mapping exercises, group discussions, and observation. This fieldwork was supplemented over a period of ten months by archival research, specifically, a study of Government Orders (GOs) relating to the Buckingham Canal dating back to the 1940s from the Tamil Nadu State Archives. We also studied secondary material such as reports and policy notes, and interviewed a cross-section of commentators, from state engineers to environmentalists and historians.

The Canal as Failed Infrastructure

The nearly 150-year-old Buckingham Canal runs down the length of Chennai city on its eastern side, in parts veering close to the city's coastline, and traversing some of its oldest and densest settlements (see Figure 9.1). The canal was built in sections between 1804 and 1883 by British colonial entrepreneurs and later the colonial government as a navigation channel to transport goods from the northern parts of the Madras Presidency (now the state of Andhra Pradesh) to the capital city, Madras (Russell 1898; Coelho et al. 2017; IWAI 2010).

Although it now runs for a length of 796 kilometers, from Peddaganjam in Andhra Pradesh to Marakkanam in Tamil Nadu, the canal is not a single coherent entity. Rather, it comprises multiple segments, with diverse landscapes, ecological features, and functions (backwater drainage, tidal balance), and varying conditions of maintenance, use, and repair (Russell 1898; Coelho et al. 2017).

The stretch within the city of Chennai is distinctive. The canal's embattled career in the city, and its ongoing struggle to survive, are telling testimonies to the city's choices and patterns of growth (see Figure 9.2). Although the canal had its genesis in the city of Madras/Chennai, its city segment has been officially declared dead for navigation purposes. In 2004 the Inland Waterways Authority of India (IWAI) proposed to revive navigation on the entire Buckingham Canal as part of National Waterway 4, to link ports and harbors along the east coast (IWAI 2010). This proposal, however, declared that the Chennai segment of the canal was not salvageable, indeed that it had "los(t) its existence" (IWAI 2010, 3). It ascribed this death partly to sewage pollution from the city, and partly to competing infrastructures, in particular the priority that had been accorded to road and rail, resulting in too many bridges over the canal within the city. The report recommended that the project bypass the canal and arrange for "multi-modal cargo transfer" within the city (IWAI 2010, 10).

As an urban waterway, the Buckingham Canal presents several ambiguities and conundra. First, although an engineered navigation channel, it is also a tidal canal intersecting and interacting with several rivers, lakes, and backwaters, and playing an important but complex part in the drainage of the eastern coastal plains (Russell 1898). Hydrologists have reported on its contribution to absorbing and mitigating the impact of cyclones, floods, and the 2004 tsunami on the city, and in channeling and draining eastward flows of monsoon waters to the sea (see Balasubramanian 2005). This has made it part of the "natural" ecology of the city, along with the city's three rivers and smaller drainage canals. However, its drainage functions have been compromised by a severe reduction in depth and width, and by encroachments of various kinds.

Second, despite its impressive length, the canal emerges—at least in Chennai city—as a low-key plebeian infrastructure. In this sense, it offers an "anti-dam" understanding of hydraulic installations, a counterpoint to the typical profile of colonial (and postcolonial) engineered water infrastructures such as dams, reservoirs, and irrigation canals. Canals, as projects of water engineering in hydraulic states like India, have been seen as grand monumental interventions into the landscape, serving a

Figure 9.1 Map of the Buckingham Canal, showing Chennai's boundaries, rivers, the MRTS, and the settlements studied

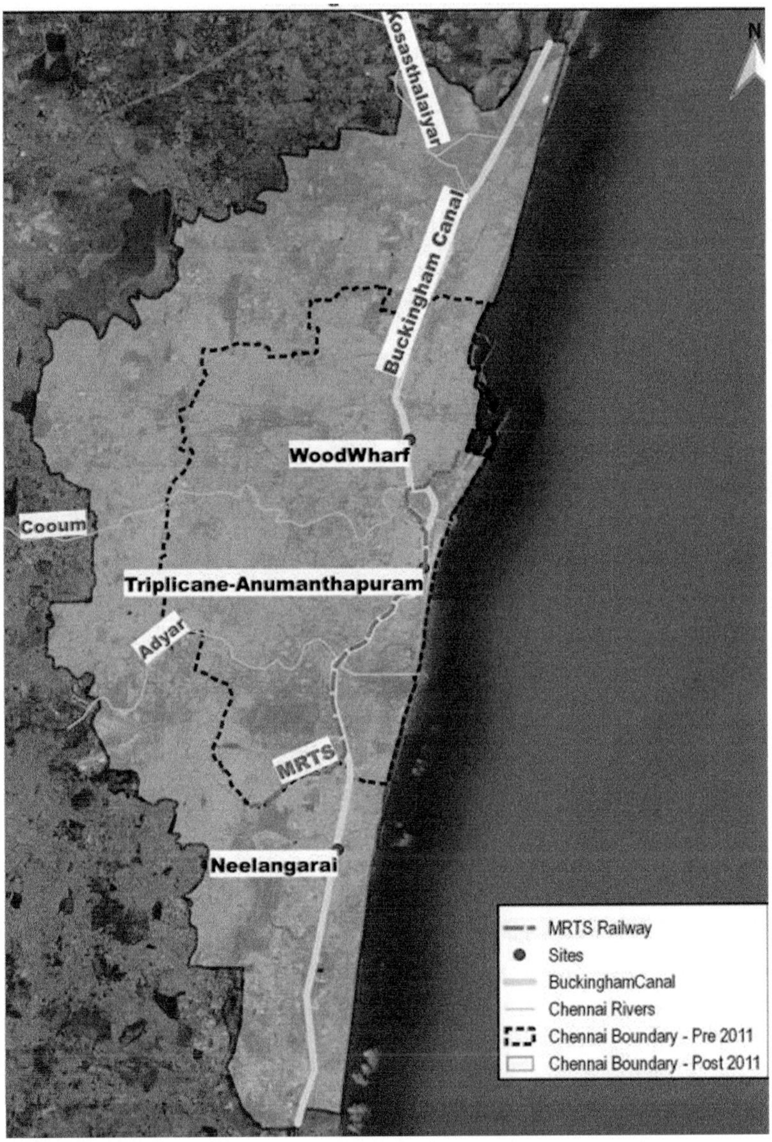

range of modernist functions, from manipulating and shaping land uses, to promoting economic activities, generating revenue streams, creating employment, and legitimating rule (D'Souza 2006; Molle et al. 2009). The Buckingham Canal has historically been seen to fulfill some of these functions. When it was completed at the end of the nineteenth century,

Figure 9.2 The Buckingham Canal running past the Central Station in Chennai (From the personal collection of D.H. Rao, used with permission)

the British engineer A.S. Russell noted that it had dramatic effects on the landscape:

> The construction of the Buckingham Canal has placed the town of Madras in cheap and easy communication with no less than five districts, and with the large and important towns of Cocanada, Bezwada, Masulipatam, Ongole and Nellore, besides numerous smaller trade centers ... It passes through what was, before its construction, a dreary waste of sand, but much of this barren and arid country has been greatly developed and improved owing to the remarkably cheap means of communication afforded by the canal; cultivation has been brought into existence or extended, owing to the facilities given by the canal for the drainage of low-lying land; numerous casuarina and other plantations have been formed along its entire length, and along the shores of backwaters in communication with it; and a great increase in the wealth and prosperity of the population has taken place. (Russell 1898, 40)

Notwithstanding this account, an overview of archival documents on the canal reveals that the Buckingham Canal was never conceived as a single grand state engineering scheme. Rather, it was patched together over about

75 years from segments of excavated channels, links, junctions, and other existing canals, built at different times by different actors, from the British entrepreneur Basil Cochrane who made the first cut in 1804, to the Public Works Department (PWD) that took it over in the mid-1800s (Rao 2015).[10] Large segments of the canal were, in fact, constructed as a food-for-work scheme, providing employment to thousands of workers after the severe famine of 1877 (Russell 1898). Once completed, the canal served to transport small wooden craft with cargoes of basic essentials such as firewood, salt, lime, and locally grown agricultural produce. While it was heavily used in certain periods and certain stretches, historical evidence suggests that its navigational function had always been under stress. Russell, in the late nineteenth century, had highlighted obstructions to its navigation caused by chronic siltation, partly from tidal action and partly from sewage discharged into the canal within the city limits. From archival accounts, it remained perennially in a state of crisis, calling for, at the minimum, frequent desilting to allow even small boats to navigate it, and at other times, thoroughgoing repairs and reconstruction (see GO 165, PWD, 19/1/1944; GO 1714, PWD, 6/4/1956).

The primary metric through which the state viewed the canal was as a commercial body. Until the late 1950s and early 1960s, the canal was still expected to yield revenue to the government, and considerable discussion occurred around strategies to enhance earnings from its operations (GO 2285, PWD, 1/9/45; GO 1299, PWD, 13/3/1956). None of these bore fruit. By the 1960s, the navigational potential of the canal was clearly on the decline. Despite periodic efforts by the PWD to deepen, widen, and desilt its bed and augment its terminal and wharfage facilities, the canal remained a challenge to maintain and operate. In 1962, the PWD wrote to the government that while the canal served to provide low-cost freight services to the public, it would never yield direct returns, as annual expenditure on its maintenance significantly exceeded the revenue earned from its operation. It argued that maintaining criteria of direct returns on investment were constraining improvement works on the canal (GO 3289, PWD, 6/12/62).

As processes of urbanization were peaking in Madras city in the 1960s, there was considerable ambivalence in official and political circles about

10 The archival documents cited here are part of an unpublished collection of archival material on the Buckingham Canal collected and curated by D.H. Rao, an amateur historian of the canal in Chennai. His collection comprises extracts of letters and notes from volumes of the Board of Revenue, Revenue Consultation, and Public Consultation, dating from 1797 to 1852, all accessed from the Tamil Nadu State Archives. Henceforth referred to as Rao, personal collection.

the functions and future of the canal in the city, including proposals to close it up and reclaim the land for other infrastructure. S. Krishnamurthy, Mayor of Madras in 1964, expressed some of this when he wrote: "it has to be considered (i) whether it would be practicable to preserve the canal entirely free from the pollution from the slums along its banks, or (ii) whether the closure of the canal and replaced (sic) by railway line would seriously affect the drainage system in the city" (Krishnamurthy and Kabalamurthy 1964, 49).

Archival documents reveal that a second major lens through which the state viewed the canal was as a chronic source of pollution, ascribed primarily to the preponderance of slums along its banks (see GO 1714, PWD, 6/4/1956; GO 3042, PWD, 17/11/1958). This view contributed to its low status in the urban landscape, and to the debates over maintaining it or closing it down.

A third significant frame in which the canal figured in the eyes of the state was as an empty tract of land available for urban infrastructure development—increasingly disconnected with, or indeed competing with, the canal's intended function of inland navigation. In this view, the canal is not so much a waterway as a convenient land corridor cutting through the city, offering a suitable alignment for the installation of rail and road links (MTS 1996).

Such re-inscriptions of water as land, or the denial of water, have indeed led to the diminishing, destruction, or disappearance of water bodies across urban centers in India. Waterlines are notoriously smudgy—on shorelines, lake edges, riverbanks, or floodplains, the demarcation of water from land is rarely a line, it is rather a shifting zone that seasonally turns into one or the other (Mathur and da Cunha 2009). But urbanization processes in India have classically proceeded through authoritatively asserting the categories of land and water on these reluctant interfaces, and reinforcing the boundaries with signboards, concrete, projects, and finance (Coelho 2020). Baviskar (2011) and Follmann (2016 and this volume), for instance, document how large tracts of the River Yamuna's floodplains in New Delhi were handed over—against the protests of environmental groups—for construction of the Commonwealth Games Village, by a Supreme Court judgement that simply declared that "the site in question is neither a 'floodplain' nor a 'riverbed'."[11] In Chennai, hundreds of seasonal waterbodies—lakes, tanks and ponds—that trapped monsoon waters, recharged aquifers, and fed the city's rivers, were

11 Supreme Court of India, 30 July 2009: Civil Appeal Nos. 4866–67 of 2009. Cited in Follman (2016, 10).

Figure 9.3 The Buckingham Canal near Triplicane, choked by rail bridges and solid waste (Photo by Karen Coelho)

eliminated from the 1960s on by large-scale government schemes. These schemes sought to enhance the supply of land for housing, infrastructure, and institutions through filling vast expanses of these waterbodies, often at considerable cost (see Coelho 2020; Coelho and Raman 2010, 2013). When one such scheme was challenged in court in 1993 by an environmental NGO, the government countered with a claim that the lakes in question were defunct (TNHB 1993). The erasures of water in the landscape are sometimes pre-emptively signaled. Jayaraman (2016) writes of a signboard standing in the middle of a marsh in Ennore in north Chennai, proclaiming "This land is the property of Kamarajar Port." Here water had to be reinvented as land in order to claim ownership over the tract.

A second related trend in contemporary Indian urbanism is the siting of infrastructure corridors such as highways and railways on urban waterways. Part of world-classing initiatives, these projects contribute to enhancing the visibility and value of the waterway, but are more mundanely motivated by the relative ease of acquisition of water's edge lands, which are typically government-owned "commons." In the early 1980s, a substantial stretch of the Buckingham Canal's banks was chosen as the alignment for the first phase of the elevated intra-city railway, the MRTS (see Figure 9.1). Executed between 1986 and 1996, this project was the final blow in the long series of abuses that the canal had suffered through its history. It is widely perceived as having contributed to the death of the canal not only for navigation, but also for drainage purposes (IWAI 2010) (see Figure 9.3). The canal alignment was selected principally because of the lower cost of land acquisition entailed (MTP 1996). Project reports of the MRTS represent the canal as a passive site, with no independent dynamism, no life, flow, or ecology. There is no concern articulated anywhere in the report about the effects of the project on the waterway (MTP 1996).

These state optics on the canal, as a non-performing asset, a chronic and intractable pollutant, or a tract of empty space, have overdetermined the way that the canal has been perceived and valued in the city's history. These utilitarian perspectives have obscured other histories of the canal, in particular, the ways in which it has configured social and economic landscapes of the city and generated a set of distinctive urbanisms closely tied to its flow through the city's history. As Hastrup (2013) notes, seeing waterbodies through the lens of hydrology, and employing "numbers and other measures of valuation … [are] ways of abstracting water from its social significance, which sometimes contribute to the fading out of the sociality of water" (2013, 59).

The Canal as a Corridor of Urban Place-Making

The edges of waterways or waterbodies, especially in low-lying or floodplain (and many coastal) cities, are typically dense settlement zones. It is often in these spaces that urban life gets seeded and formed for working-class migrant families, for various reasons. The water's edge provides many of the bare needs of the migrant's life—food, wild or cultivated, space to pitch shacks, material to build them, water and drainage. These are open lands—wild, uninhabited, and considered unsuitable for settlement by any but the most desperate, who then expend considerable effort in taming the lands to their needs. They level the ground, clear undergrowth, fight off snakes, and concretize the slush and mud into homes, workshops, and temples.[12] But these lands are also "open" in revenue terms: as *poramboke* lands, they are unavailable for sale or purchase, hence avoided by elite settlers in pursuit of secure property. Not surprisingly, then, water's edges in old cities have historically evolved as spaces of informal settlement.

12 Oral histories from Coelho et al. (2017).

But, as we learned from the Buckingham Canal, such processes of urbanizing do not yield linear historical narratives of greenfield sites being transformed into stable urban neighborhoods. Rather there is a reiterative cycle of production, appropriation, and destruction, of settling, uprooting and moving, resettling, being resettled. Urban river- and canal banks and lakeshores are zones of contestation, between their typical histories of squatting and auto-construction, multiple other visions, plans and schemes, and failures of implementation and governance that are played out along these borderlands (Baviskar 2011; Mathur 2012; Coelho and Raman 2013; Coelho 2020).

From the early twentieth century, the Buckingham Canal and the economies associated with its operations became a significant draw for informal settlement, and by mid-century its banks were host to large concentrations of slums. In the northern sections of the canal in the city, wharves, landing docks, and warehouses offered an abundance of manual jobs as loaders, boatmen, lascars, and watchmen, that attracted migrant workers from different parts of the state. Further south, the uninhabited, forested lands on the canal banks offered dwelling spaces for subaltern communities that often settled in groups, clearing the land and establishing small informal livelihoods such as cattle-rearing, firewood sales, sand-panning, local trades, and vending. With the spurt in migration into the city in the 1950s and 1960s, the entire length of the canal banks offered settling space for working-class families who came to work in the expanding industrial sector of Madras. By 1960, slums along the Buckingham Canal accounted for about 10 percent of the total number of recorded slums in the city, housing almost 10,000 families (Census of India 1961).

Our ethnography on the canal banks (Coelho et al. 2017) examined how the water's edge had spawned and shaped the emergence of distinctive localized urbanisms in each of the three sites studied. The flow of the canal was not simply a backdrop to these developments, nor was it a homogenous phenomenon across the three settlements. Instead the particular historical, ecological, and urban features of the canal emerged as part of the story of place-making in each site.

The settlement of Wood Wharf in the northern part of the city emerged as a direct consequence of the livelihoods provided by the Buckingham Canal. Wharves had been built here in the early 1900s to offload firewood, salt, dried fish, and other material transported from northern parts of the Madras Presidency. Families, mostly *dalit*, who migrated from nearby districts to work on the wharves, pitched tents or built huts on open lands slightly inland from the canal. When navigation began to decline in the canal, the PWD

Figure 9.4 Warehouses and godowns along the canal at Wood Wharf, 1970s (From the personal collection of D.H. Rao, used with permission)

resorted to leasing out lands on the canal banks to supplement revenues. Individuals and business houses took the leases and constructed godowns, sheds, and warehouses, which changed hands and uses over time (GO 1669, PWD, 9/6/1958) (see Figure 9.4). Oral histories from our study revealed that by the 1970s, with capital from the Marwari trading community settled nearby, these godowns and sheds were converted into small manufacturing units that fabricated a wide range of aluminum and stainless steel products, from plates and utensils to large drums and metal trunks for sale in local wholesale markets and neighboring states (see Figure 9.5). Other small-scale works such as cleaning and recycling of large plastic drums were also established here. Over the past 40 years, these lands and sheds have morphed into a vibrant and buzzing small-scale manufacturing hub with intense multipliers and linkages. It directly employs thousands of unskilled and semi-skilled workers, both from the neighborhood and from other parts of north Chennai, and indirectly provides livelihoods to hundreds of local families through a proliferating support economy of logistics, maintenance, water supply, and canteens.

The residential plots that were established among these industrial units and slightly further inland never obtained formal titles as the land belonged to the PWD. Consequently, very few families have ventured to

Figure 9.5 A steel workshop in Wood Wharf (Photo by Stephen Amirtharaj, used with permission)

build upward, and living conditions in the area have become increasingly congested. The population has increased exponentially. As one resident observed, "Few people move out and more keep coming." Wood Wharf displayed many features of a classic Third World "slum,"[13] with little or no sanitation facilities, homes set among noisy lathe workshops and industrial godowns along a highly polluted canal, and high rates of male alcoholism. Yet residents embraced this modernity of dense agglomeration, diversity, and opportunity, where Marwari businessmen operated side-by-side with Christian evangelists, and daily wage workers educated their children in English if possible.

Respondents insisted that this was a place that had provided for them across two or three generations, and had allowed them access to education and economic mobility. They found it unthinkable to move elsewhere. Several residents owned flats in other parts of the city, but would not leave this place despite the squalor and congestion. The precarity of settlement here was highlighted during our fieldwork when a section of families that lived on the edge of the canal was evicted. This provoked discussions on the possibilities of the larger settlement being removed and resettled outside

13 The term is employed here to index a trope rather than an objective category. For discussions on the problematics of the term see Gilbert (2007).

the city, a prospect that was strongly rejected by all our respondents. Yet, struggles for formal titles to land and houses were not a preoccupation here, possibly because residents believed they had little chance of obtaining them. Here then, the urbanism engendered by the canal's historical operations was one of informal industrial and entrepreneurial livelihoods, which sustained a strong sense of local belonging despite the low residential values.

The second site, in the central canal section of Triplicane/Mylapore, comprised a string of small, spatially contiguous yet somewhat distinct neighborhoods, housing a historically and culturally diverse set of communities, mostly of Hindu *dalits*, Muslims, or low-caste Christians. The urbanisms found here are rooted in a history of intensive slum growth along the banks of the waterway from the 1920s on, and subsequent struggles over state slum clearance schemes. It was this stretch of the Buckingham Canal, between the rivers Cooum and Adyar, that gave the canal its reputation as a corridor of slums (Census of India 1961), and it was here that the Tamil Nadu Slum Clearance Board, established in 1971, carried out one of its first detailed slum surveys. In 1975, the Board drew up a special scheme, the Buckingham Canal Scheme, to clear 40 slums and rehouse over 6000 families in state-built tenements in situ, testifying to an era when the government was willing to allocate center-city lands to accommodate working-class families (GO 1004, HUDD, 17/7/1976).

However, these plans did not materialize as envisioned. Many of these communities rejected the tenement schemes and instead demanded titles for the lands they occupied (GO 1129, HUDD, 30/8/1976). At the time of our study, the planned tenements were in evidence in only eight of the 40 slums named in the proposal. Even as the tenement scheme was being rolled out in the late 1970s, an official shift in slum clearance policies interrupted and diverted its implementation. The World Bank's entry in 1972 into the urban housing sector of Tamil Nadu and its substantial funding support for the Madras Urban Development Projects (MUDP) I and II (implemented between 1977 and 1988) provoked a shift in official approaches from expensive tenement construction to in-situ slum upgrading and Sites and Services schemes, as part of an emerging global consensus on best practices (Raman 2011). Many of the canal-bank communities slated for tenement accommodation demanded to be included in the in-situ improvement schemes and succeeded in retaining their plots (GO 1129, HUDD, 30/8/1976). Through decades of incremental improvements, these settlements have morphed into solid urban neighborhoods of two- and three-storied buildings, brightly painted and densely occupied, providing some of the largest stocks of affordable housing in the central part of the

Figure 9.6 Auto-constructed housing along the canal in Triplicane, as viewed from an MRTS train running along the canal (Photo by M. Subadevan, used with permission)

city (see Figure 9.6). Here, the 1960s generation of unskilled manual laborers have found socioeconomic mobility, educated their children, and integrated themselves into the white-collar service economies of the area (Coelho et al. 2017).

The third settlement, Arignar Anna Nagar in Neelangarai, is located on the southern stretches of the canal as it leaves the city running south. This settlement represents a second-generation squatting, wherein families already settled in the city, often as renters, moved out in the 1980s in search of open lands that they could own as property (Coelho et al. 2017). The squatting here was not aimed at acquiring shelter or a foothold in the city's livelihood landscape; rather, urban footholds already gained were surrendered in favor of entering the increasingly dominant mode of income-generation through rents. Respondents recalled their southward migration down the canal as a move from urban to rural lifestyles. Yet it also involved a re-enactment of the squatter cycle of taming wild lands, erecting shacks, and building bridges and bunds across and along the canal (see Figure 9.7). In short, we track here the transformation of the canal banks into an up-and-coming urban neighborhood in the span of three decades through classic "occupancy urbanism" processes (Benjamin 2008) of working-class settlers incrementally

Figure 9.7 Arignar Anna Nagar, Neelangarai settling and building in the 1980s (Source: Neelangarai resident's archive, used with permission)

and collectively building up essential infrastructure from scratch, in the quest of becoming land owners.

Despite the fact that none of the residents have obtained land titles, the settlement today is a far cry from a slum. It comprises a neat row of brick-and-concrete houses inside walled yards, many built up with an extra floor, with street lights, water facilities, small grocery stores, and several temples along a paved road running beside a quietly flowing, if highly polluted canal (see Figure 9.8). Settlers here came from diverse religious and caste groups, rendering the community very diverse—a quality that has been enhanced since 2000 with the arrival of a new wave of migrants from the northern and northeastern states of India who rent rooms from the locals. Rental incomes are a significant source of household income here. Access to education and employment have improved, and several of the youth worked in skilled white-collar jobs along the nearby IT corridor. Most residents of Arignar Anna Nagar claimed that they had few problems any more apart from the toxic state of the canal and the insecurity of their tenure. However, aspirations of residents here centered heavily on obtaining titles, which, they felt, would bestow respectability and decency on the neighborhood, and would propel the area into multistoried development with expanded potential for rentals.

Figure 9.8 Houses along the canal in Arignar Anna Nagar, Neelangarai (Photo by Karen Coelho)

What does the canal mean today to the communities settled on its banks? How have the socio-spatial transformations on the water's edge impacted their relations with the waterway? We found that perspectives on the canal had changed over time even for the subaltern settlers. As the water darkened with ever heavier loads of urban pollution, the settlements turned their backs on it. Although each of the three settlements had arisen out of different relations with the waterway and were embedded in different ecologies of the canal, they all evinced this switch in perspective. Even in the older neighborhood of Wood Wharf, born directly of the canal's operations, its frontal relations with the water had withered. The canal was now a rear, the space for its effluents and wastes (see Figure 9.9). In the central and southern areas, the canal that had once attracted settlers and provided them with the basic conditions of sustenance, had become no more than a nuisance. Some older residents shared nostalgic recollections of days when they played, traveled, and bathed on the canal. But today it is a sore, a source of pollution and ill-health, the most marked feature of their slum-like conditions. In some areas, the most vulnerable newcomers built their houses there, and tried to shield themselves somehow from the murky waters lapping at the garbage-strewn banks. Settlers on the extreme edge remain the most susceptible to eviction.

Figure 9.9 The canal as backyard (Photo by Karen Coelho)

Conclusions

What do we gain from this framing of water's edge urbanisms as a category of intra-urban peripheral formation? First, such a framing draws attention to dynamic but precarious projects of urbanization that unfold along urban waterways, shaped by the changing political ecologies of the water's edge. In tracking the shifting fortunes of the Buckingham Canal, this chapter reveals how its genesis as an engineered navigation canal and its declining career over the late nineteenth and early twentieth centuries were strongly influenced by utilitarian state metrics of revenue generation and infrastructure building. If state perspectives judged the canal a failure in its hydrological and infrastructural functions, we argue that in its social landscapes and in the distinctive economies that it has spawned along its length, the canal has built a productive and enduring legacy for Chennai's formation and growth. Annual celebrations of the Buckingham Canal as part of Chennai's urban heritage overlook the histories of opportunity, mobility, and transformation that its edge spaces have afforded to subaltern communities of the city.

Second, analyzing this "periphery" through the UPE lens urges us to reframe urban ecologies as complex socio-natural systems in which the waterway is closely linked with the urban social formations that have surrounded and co-constituted it. The revalorizing of water that has occurred at specific junctures in the city's history through eco-restoration or waterfront development projects has been accompanied by a punitive devalorization of these social-historical offshoots of urban water. The politics of eco-restoration in Indian cities tends to pit environmental values against those of social inclusion. This chapter seeks to dismantle that dichotomy and dislodge purist and ahistorical perspectives of waterways as pristine nature that can and should be rescued from the polluting urban social (see also Coelho 2020). It seeks to encourage a more expansive imagination wherein the long-established socio-natural relations embedded in water's edge urbanisms can be mobilized to restore and revitalize water in the city.

And third, the corridor character of these water's edge peripheries, whether sea- or lakeshores, river- or canal banks, signal a politics of scale and connection. Forging common stakes among water's edge communities could not only challenge the scapegoating of local communities for pollution generated upriver, but could possibly open up a repertoire of strategies to confront the various threats that the water's edge encounters.

References

Balasubramanian, D. 2005. "Buckingham Canal Buffered Tsunami Fury." *The Hindu*, 11 August. Available at http://www.thehindu.com/seta/2005/08/11/stories/2005081100171500.htm, accessed 16 December 2021.

Baviskar, Amita. 2002 "The politics of the city', *Seminar*, 516: 40–42.

Baviskar, Amita. 2011. "What the Eye Does Not See: The Yamuna in the Imagination of Delhi." *Economic and Political Weekly, Review of Urban Affairs*, 46(50): 45–53.

Benjamin, Solomon. 2008. "Occupancy Urbanism: Radicalizing Politics and Economy beyond Policy and Programs." *International Journal of Urban and Regional Research*, 32(3): 719–729.

Caldeira, Teresa. 2016. "Peripheral Urbanization: Autoconstruction, Transversal Logics, and Politics in Cities of the Global South." *Environment and Planning D: Society and Space*, 1–18. https://doi.org/10.1177/0263775816658479.

Census of India. 1961. Vol IX. Part XI-C: *Slums of Madras City.*

Coelho, Karen. 2017. "The Canal and the City: An Urban-Ecological Lens on Chennai's Growth." *India International Center Quarterly, Special Issue on The Contemporary Urban Conundrum*, Winter 2016–Spring 2017: 87–300.

Coelho, Karen. 2018. "Reading History and Power in Urban Landscapes: The Lens of Urban Political Ecology." *Ecology, Economy and Society—The INSEE Journal*, 1(2): 19–30.

Coelho, Karen. 2020. "Lines in the Mud: Tank Eco-Restoration and Boundary Contestations in Chennai." *Urbanisation*, 5(2): 121–139.

Coelho, Karen, Anusha Hariharan, Meghna Sukumar, and T. Venkat. 2017. "Settlement and Struggle on Chennai's Buckingham Canal: Working Class Histories of the City." *A Case Study for Reframing Urban Inclusion: Teaching and Learning Cases from India and the Global South*, curated by Indian Institute for Human Settlements. Available at http://cases.iihs.co.in/#casearchive/-KW1dn-bB93VvMVD2e2r, accessed 16 December 2021.

Coelho, Karen, and Nithya Raman. 2010. "Salvaging and Scapegoating: Slum Eviction on Chennai's Waterways." *Economic and Political Weekly*, 45(2): 19–23.

Coelho, Karen, and Nithya V. Raman. 2013. "From the Frying-Pan to the Floodplain: Negotiating Land, Water and Fire in Chennai's Development." In *Ecologies of Urbanism in India: Metropolitan Civility and Sustainability*, edited by Anne Rademacher and K. Sivaramakrishnan, 145–168. Hong Kong: Hong Kong University Press.

D'Souza, Rohan. 2006. *Drowned and Dammed: Colonial Capitalism and Flood Control in Eastern India*. New Delhi: Oxford University Press.

Follman, Alexander. 2016. *Governing Riverscapes: Urban Environmental Change along the River Yamuna in Delhi*. Stuttgart: Franz Steiner Verlag.

Friedmann, John. 2007. "Reflections on Place and Place-Making in the Cities of China." *International Journal of Urban and Regional Research*, 31(2): 257–279.

Ghertner, Asher. 2015. *Rule by Aesthetics: World-Class City Making in Delhi*. New York, NY: Oxford University Press.

Gilbert, Alan. 2007. "The Return of the Slum: Does Language Matter?" *International Journal of Urban and Regional Research*, 31(4): 697–713.

Goldman, Michael, and Devika Narayan. 2019. "Water Crisis through the Analytic of Urban Transformation: An Analysis of Bangalore's Hydrosocial Regimes." *Water International*, 44(2): 95–114.

Hastrup, Kirsten. 2013. "Water and the Configuration of Social Worlds: An Anthropological Perspective." *Journal of Water Resource and Protection*, 5: 59–66.

Heynen, Nik. 2014. "Urban Political Ecology I: The Urban Century." *Progress in Human Geography*, 38(4): 598–604.

Inland Waterways Authority of India (IWAI). 2010. *Final DPR for Development of Navigation in Kakinada-Puducherry Canal along with Rivers Godavari and Krishna*. Gurugram: WAPCOS Limited.

Jayaraman, Nityanand. 2016. "Chennai May Just Be Masterminding Its Next Flooding Disaster." *The Wire*, 21 June 2016. Available at https://thewire.in/44393/chennai-may-just-be- masterminding-its-next-flooding-disaster, accessed 10 January 2021.

Krishnamurthy, S. (Mayor of Madras), and A.A. Kabalamurthy (Deputy Mayor). 1964. *Greater Madras*. Unpublished monograph.

Mathur, Anuradha, and Dilip da Cunha. 2009. *Soak: Mumbai in an Estuary*. Mumbai: National Gallery of Modern Art/Rupa and Co.

Mathur, Navdeep. 2012. "On the Sabarmati Riverfront: Urban Planning as Totalitarian Governance in Ahmedabad." *Economic and Political Weekly*, 47(47–48): 64–75.

McDuie-Ra, Duncan. 2012. "The 'North-East' Map of Delhi." *Economic and Political Weekly*, 47(30): 69–77.

Metropolitan Transport Project (Railways) Madras (MTP). 1996. *Mass Rapid Transit System: Madras City-Phase II.*

Molle, Francois, Peter P. Mollinga, and Philippus Wester. 2009. "Hydraulic Bureaucracies and the Hydraulic Mission: Flows of Water, Flows of Power." *Water Alternatives*, 2(3): 328–349.

Mukerji, Chandra. 2009. *Impossible Engineering: Technology and Territoriality on the Canal du Midi.* Princeton, NJ: Princeton University Press.

Natarajan, A. 2018. "Winning Ideas to Revive Buckingham Canal and Make Chennai Climate-Proof: Eyes on the Canal Competition Results." *Citizen Matters*, 10 December 2018.

Padawangi, Rita. 2019. "Forced Evictions, Spatial (Un)certainties and the Making of Exemplary Centres in Indonesia." *AsiaPacific Viewpoint, Special Issue: The Production and Contestation of Exemplary Centres in Southeast Asia*, 60(1): 65–79.

Rademacher, Anne M. 2008. "Fluid City, Solid State: Urban Environmental Territory in a State of Emergency, Kathmandu." *City and Society*, 20(1): 105–129.

Raman, Nithya. 2011. "The Board and the Bank: Changing Policies towards Slums in Chennai." *Economic and Political Weekly*, 46 (31): 74–80.

Rao, D. Hemachandra. 2015. "How the Buckingham Canal Was Born." *Madras Musings*, 24(19): 16–31.

Russell, A.S. 1898. *History of the Buckingham Canal Project, with a Descriptive Account of the Canal and Its Principal Works, and a Guide to Its Future Maintenance.* Compiled for the Government of Madras.

Shannon, Kelly. 2013. "Eco-engineering for Water: From Soft to Hard and Back." In *Resilience in Ecology and Urban Design: Linking Theory and Practice for Sustainable Cities*. Future City Series, edited by S.T.A. Pickett, M.L. Cadenasso, and Brian McGrath, 163–182. New York, NY and London: Springer.

Simone, AbdouMaliq. 2004. *For the City Yet to Come: Changing African Life in Four Cities.* Durham, NC: Duke University Press.

Simone, AbdouMaliq. 2017. "Living as Logistics: Tenuous Struggles in the Remaking of Collective Urban Life." In *The Routledge Companion to Planning in the Global South*, edited by Gautam Bhan, Smita Srinivas, and Vanessa Watson, 264–275. Abingdon-on-Thames: Routledge.

Tamil Nadu Housing Board (TNHB). 1993. Counter-affidavits of the Tamil Nadu Housing Board (TNHB) and Housing Department, Government of Tamil Nadu in W.P. No. 17914/93 before the Madras High Court—Consumer Action Group vs.

Project Director, Project Management Group, Chairman, Tamil Nadu Housing Board and Six Others.

Venkateshwarlu, K. 2017. "An Ambitious Plan to Revive the Buckingham Canal." *The Hindu*, 5 August 2017. Available at https://www.thehindu.com/news/national/andhra-pradesh/long-live-the-canal/article19429503.ece, accessed 16 December 2021.

Wade, Robert. 1985. "On the Sociology of Irrigation: How Do We Know the Truth about Canal Performance?" *Agricultural Administration*, 19(2): 63–79.

Archival Sources Cited

GO 165, Public Works Department, 19/1/1944. "Madras City – Sources of Pollution of Waterways – Remedial Measures – Joint Report of the Sanitary Engineer and Superintending Engineer, Madras Circle – Communicated to the Corporation of Madras." Accessed from the Tamil Nadu State Archives.

GO 1004, Housing and Urban Development Department, 17/7/1976. "Slum Clearance Scheme – Madras City. Remark of Slums on the Banks of South Buckingham Canal. Construction of Tenements for Slum Dwellers under 2 Phases – Proposal of S.C.B. Approval." Accessed from the Tamil Nadu State Archives.

GO 1129, Housing and Urban Development Department, 30/8/1976. "MMDA Madras – Madras Urban Development Project – World Bank Appraisal 8.1 Component Orders Approving the Proposal – Issued – Paper Re-closed." Accessed from the Tamil Nadu State Archives.

GO 1299, Public Works Department, 13/3/1956. "Possibilities of Curtailment of Expenditure on Buckingham Canal." Accessed from the Tamil Nadu State Archives.

GO 1669, Public Works Department, 9/6/1958. "Land – Buckingham Canal – Wood Wharf Area – Lease to Corporation of Madras – Collection of Rent – Appeal – Orders Passed." Accessed from the Tamil Nadu State Archives.

GO 1714, Public Works Department, 6/4/1956. "Periodical Cleaning of Cooum and Buckingham Canal." Accessed from the Tamil Nadu State Archives.

GO 2285, Public Works Department, 1/9/1945. "Buckingham Canal – Transport – Speeding Up of Traffic – Hiring Out Government Boats to Merchants – Debit of Loss – Ordered." Accessed from the Tamil Nadu State Archives.

GO 3042, Public Works Department, 17/11/1958. "Buckingham Canal – Nuisance from Bad Smell – Steps Taken – Papers Recorded." Accessed from the Tamil Nadu State Archives.

GO 3289, Public Works Department, 6/12/1962. "Working of Buckingham Canal and Vedaranyan Canal. Papers Recorded." Accessed from the Tamil Nadu State Archives.

Biography

Karen Coelho is an associate professor at the Madras Institute of Development Studies in Chennai, India. She received her PhD in anthropology from the University of Arizona at Tucson in 2004. Her work includes studies of informal settlements and urban ecologies, and critical examinations of urban collective action and urban infrastructure including housing and water.

10 The Social Downside of Flood Control

The Case of a Marginalized Community in Taipei

Kuei-Hsien Liao

Abstract

Flood control infrastructure confers safety for some but transfers floodwater and associated flood risk to others who are already socioeconomically marginalized. This downside of flood control is demonstrated by Shezidao in Taipei. While the socioeconomically more important areas of Taipei are protected against a 200-year flood, Shezidao has been defined as a "flood zone" and its physical development legally prohibited since 1970. As a result, Shezidao has experienced the deterioration of its living environment. Shezidao's residents have felt a sentiment of injustice that they are being sacrificed for the flood safety of the rest of the city. This chapter discusses the important issue of environmental justice associated with flood control as the dominant approach to flood management in river cities around the world.

Keywords: environmental justice, flood control infrastructure, floodwater redistribution, flood risk transfer, Shezidao, Taipei

Introduction

Cities that originated along rivers used to depend heavily on the rivers for water supply, navigation, and fish as food. However, the natural phenomenon of riverine flooding is often considered a hazard that would impede urban development. Flood control—the deployment of river engineering works to prevent riverine flooding—has been the dominant approach to flood management. The ubiquity of flood control infrastructure (FCI)—here referred to as the collection of structural measures such as levees, floodwalls, channelization, flood diversion—in river cities around the world bears

Padawangi, Rita, Paul Rabé and Adrian Perkasa, *River Cities in Asia*.
Taylor & Francis Group, 2022
DOI: 10.5117/9789463721851_CH10

testimony to the prevailing ideology that flood control is necessary for the development of modern urbanism.

While FCI can significantly reduce the frequency of flooding, it has serious side effects. By drastically altering the geomorphology and hydrology of rivers, FCI has led to the severe ecological destruction of riverine ecosystems around the world (Postel and Richter 2007). Moreover, the sacrifice of ecological health does not deliver long-term flood safety. FCI simply cannot prevent all floods because FCI is limited by the design standard, and hence, residual risk still exists (Hewitt and Burton 1971). To better manage flood risks, nonstructural measures, such as land use regulation, flood warning, and flood insurance, have been added to the flood management toolbox. While the aforementioned side effects of FCI have been well recognized and somewhat addressed, the negative social impacts of FCI have yet to be explicitly confronted. It is widely known that large dams that control flooding for downstream urban centers have displaced numerous people upstream (Tilt et al. 2009); however, the social impacts of other structural measures are much less well documented. Available literature has suggested that protecting some groups with FCI can end up negatively affecting other groups (e.g., Douglas et al. 2009; Padawangi and Douglass 2015; Pinter et al. 2016; van Voorst and Hellman 2015). This downside of FCI can be considered a form of environmental injustice but has so far been ignored in decision making (Liao et al. 2019).

FCI can lead to environmental injustice because of its very purpose of flood prevention, as FCI prevents flooding of a particular area simply by channeling floodwater elsewhere (Liao et al. 2019). The most commonly seen structural measures in river cities are channelization, levees, and floodwalls (Brookes 1988). In effect, they modify flood frequency by altering the spatial and temporal distribution of river discharge. Channelization is used to increase the hydraulic efficiency of the river, thereby redistributing floodwater from upstream to downstream by draining the flow faster. Levees and floodwalls redistribute floodwater from upstream to downstream by preventing floodwater from spreading over the river's natural floodplain leading to greater flow magnitude downstream. When the channel is clogged or during high tides, levees and floodwalls can cause the flow—which can no longer spread over the floodplain—to back up and flood the upstream area; in this way, they can also redistribute floodwater from downstream to upstream. When floodwater is redistributed to places inhabited by people, it increases their flood risks. This is why Lebel and Sinh (2009) argue that the so-called flood risk "reduction" is often simply flood risk "redistribution."

Using FCI to provide flood safety for some people is often at the expense of others. This is especially obvious in watersheds where communities are protected by FCI to different degrees, as defined by different design standards of the structural measures. Unequal—selective—protection in a watershed subjects unprotected communities and communities with less protection to increased flood risks. Nevertheless, this problem cannot be resolved by building more FCI to equalize flood protection. Since FCI redistributes floodwater and subsequently transfers flood risk, building FCI for one place usually necessitates more FCI to be built to protect other places immediately upstream or downstream (Pinter et al. 2016; Tobin 1995). This leads to a vicious cycle whereby FCI continues to be needed (Liao 2014). In this chapter, I demonstrate the conundrum of unequal protection in a marginalized community, Shezidao in Taipei City, the capital city of Taiwan, which has been further marginalized by the unequal protection of FCI.

When resources are limited, which communities should be protected by FCI first? To what degree should a community be protected? These questions are really more sociopolitical than technical, involving a value judgment that entails ethics—who should be protected, and at whose expense. The problem of floodwater redistribution and flood risk transfer is akin to the siting controversies, where it is often the socioeconomically marginalized communities, on which NIMBY (not-in-my-back-yard) facilities are imposed (Mata 1993; Taylor 2014). Hazardous floodwater—along with flood risk—is often shifted by FCI to the same type of communities, as cost-benefit analysis is widely used to determine who should be protected by FCI and to what degree. Therefore, the economically—often also socially—marginalized communities frequently end up bearing the costs of the transferred flood risk by FCI that protects the economically more important communities (Douglas et al. 2009; Lebel and Sinh 2009). Shezidao is just one case in point. Nevertheless, those who are protected by and have benefited from FCI rarely recognize this moral implication.

The highly emphasized economic benefits of FCI have overshadowed its many pitfalls. Therefore, this case study on Shezidao aims to shed light on the social downside of FCI, which has yet to receive adequate attention. This case study relies mainly on archival research, observational studies, and ethnographic interviews. The analyzed archives include government plans, reports, and minutes from more than 20 review committee meetings and town hall meetings on the current redevelopment plan of Shezidao. Numerous observational studies have been conducted since August 2014, including attendance in 14 of the aforementioned meetings, the latest of which was in September 2020. Ethnographic interviews with local residents were conducted between February 2015 and September 2020. Furthermore, gray literature was also consulted, including locally published books and master theses, and news reports.

The chapter is organized as follows. The second section provides the important background of Shezidao as a "flood zone," and thus a "development restricted zone," and how Shezidao is related to the FCI that protects the Taipei Area. The third section reports on the difficult living conditions in Shezidao as a development restricted zone. The fourth section recounts the many attempts to "develop Shezidao" as a solution to its plight and how the FCI of the Taipei Area has played an impactful role. The fifth section further elaborates the issue of environmental injustice associated with FCI. I conclude the chapter by suggesting a paradigm shift in flood management to flood adaptation to address the social downside of flood control.

Shezidao, a "Flood Zone"

Shezidao (社子島) is located at the western border of Taipei City (Figure 10.1). With a total area of 302 hectares, it is positioned at the confluence of the Danshui River and the Keelung River, two major rivers in northern Taiwan. While in Mandarin the term "dao" (島) means "island," today Shezidao is not the island it used to be. The realignment of the Keelung River in 1965 and the filling of the waterway that used to separate Shezidao from the rest of the city in 1975 have made Shezidao a peninsula today, connected to the city by a narrow belt. As of 2015, the population of Shezidao is 11,135, with 4228 households (Taipei City Government 2018).

With an average elevation of 2.5 meters above sea level, Shezidao is essentially a sand bar formed by two rivers. It was historically a shifting terrain, extremely prone to natural flooding (Wang 2001). Few historical documents have been preserved to tell the detailed history of Shezidao, but remaining evidence suggests that farming activities appeared in Shezidao as early as the early eighteenth century (Wang 2001). However, it was not until the early twentieth century when settlements started to form in Shezidao, after the sand bar became more stable following river engineering works elsewhere in the watershed (Lin 2009). The settlements in Shezidao are located along the relatively higher ground at the bank of Danshui River (Figure 10.2). In the 1960s and 1970s, Shezidao was a major supplier of vegetables to Taipei City. Compared to the earlier lowland settlements in the southeast of Taipei City that have long been turned into burgeoning urban districts, the settlements in Shezidao are not only geographically, but also socioeconomically

Figure 10.1 Map showing the location of Shezidao (Map adapted by Yin-Ling Huang)

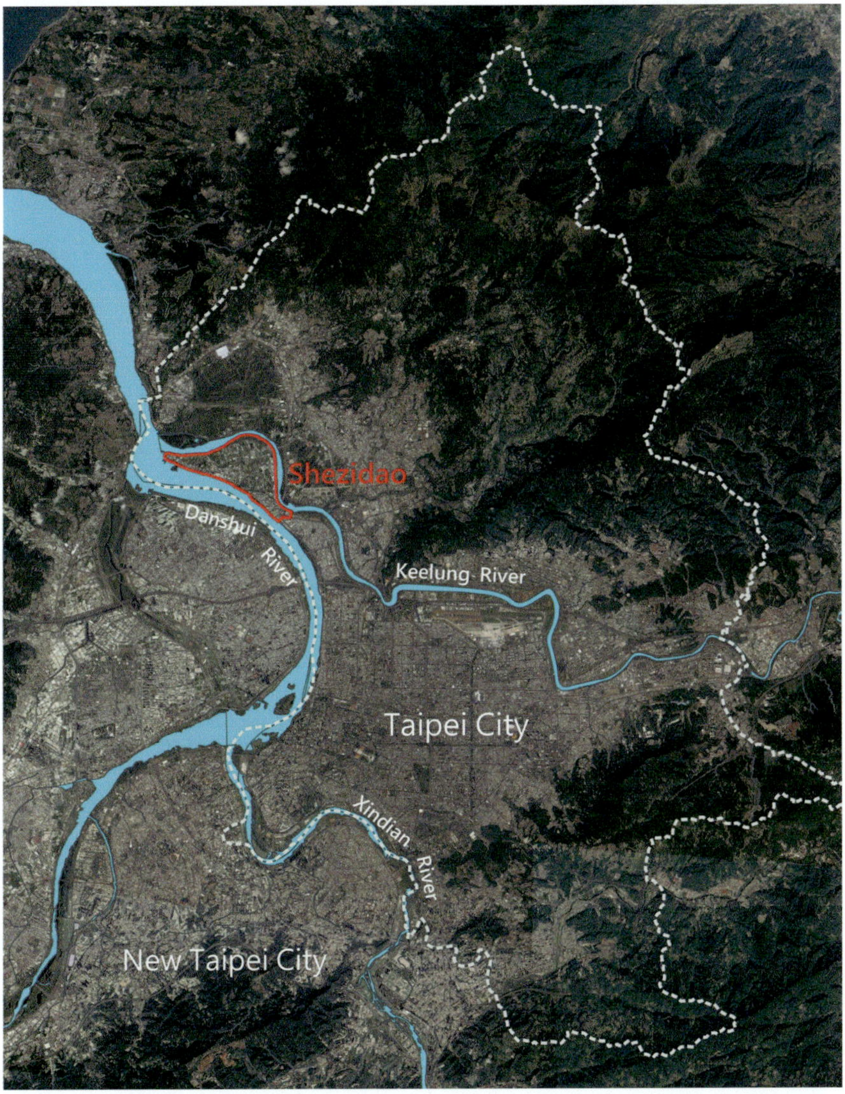

Shezidao (red line) is an area of Taipei City (dashed line) within the larger Taipei Basin (the built-up area in the aerial photo). This low-lying basin is often referred to as the Taipei Area and includes some parts of Taipei City and New Taipei City.

marginal in Taipei City. Today, Shezidao is even more marginalized, largely because of the Taipei Area Flood Control Plan (TAFCP), a large-scale capital project that involved numerous structural measures that formed the FCI of the Taipei Area.

Figure 10.2 Map showing major settlements in Shezidao (Map adapted by Yin-Ling Huang)

Most of the major settlements in Shezidao (shown in brown) are located at the relatively higher ground along the bank of Danshui River. The Yanping North Road is the main thoroughfare in Shezidao. Shezidao is currently protected by a levee system against a 20-year flood, while the rest of Taipei City is protected against a 200-year flood. At the narrow belt, where Shezidao is connected to the rest of the city, a sluice gate was proposed to protect the rest of Taipei City during extreme events. The sluice gate was eventually not constructed.

Shezidao and the Taipei Area Flood Control Plan (TAFCP)

The TAFCP was officially launched by the Central Government of Taiwan in 1973, with the objective to protect the Taipei Area against a 200-year flood (i.e., a flood with a 0.5 percent probability of occurrence in any given year). The Taipei Area is the most important economic, political, and cultural center in Taiwan. It includes the majority of Taipei City, the capital city of Taiwan, and some of New Taipei City, the most populated city in Taiwan (Figure 10.1). It is geographically defined by the Taipei Basin and drained by the Danshui River and its major tributaries. The main thrust of the TAFCP is levees and floodwalls, in conjunction with the Erchung Floodway to confine and reduce flood flows in the Danshui River and its major tributaries (Figure 10.3). Besides diverting excess floodwater during emergency events, the Erchung Floodway also functions as a park. Other measures in the TAFCP include channelization, urban drainage systems, and pumping stations.

Figure 10.3 Overview of the FCI of the Taipei Area (Map adapted by Yin-Ling Huang)

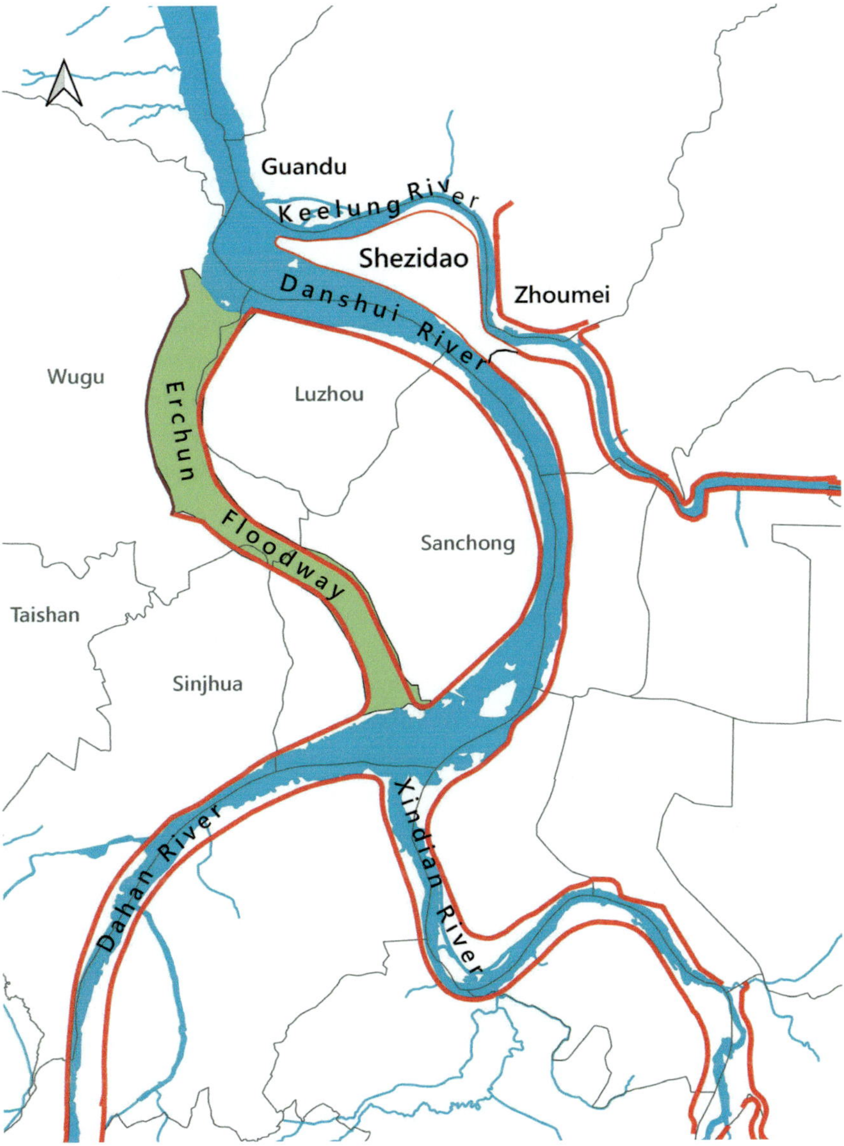

The Danshui River has three major tributaries: Keelung River, Xindian River, and Dahan River. The TAFCP provides protection against a 200-year flood for the Taipei Area, except for Shezidao, Guandu, and Wugu. The TAFCP relies on levees and floodwalls (shown as thick red lines), in conjunction with the Erchung Floodway (shown in green), which also functions as a park and is used for flood diversion only during emergency events to reduce flood flows in the Danshui River and its major tributaries. Shezidao is currently protected by a levee system (shown as the thin red line) against a 20-year flood.

The construction of the TAFCP spanned almost two decades from 1981 to 1999. Between 2001 and 2005, the Central Government implemented the Comprehensive Keelung River Improvement Plan (CKRIP), another capital project to address the upstream sections of the Keelung River that were not included in the TAFCP. The objective is to protect this relatively recently urbanized area along the Keelung River against a 200-year flood by levees, channelization, and flood diversion. The TAFCP and CKRIP together formed the FCI for the Taipei Area. Today the TAFCP not only refers to the entire FCI of the Taipei Area, but also to the capital plan itself, depending on the context.

After decades of flood control works, today the Taipei Area is theoretically capable of defending against a 200-year flood, except for Shezidao, a portion of Guandu, and a small part of Wugu (Figure 10.3). Wugu is an area along the left bank of the Erchung Floodway. Guandu is sparsely populated and largely a nature preserve of wetlands north of the Keelung River. These three places were rural in the 1970s when the TAFCP was proposed and were all excluded from the plan.

The TAFCP's impact on Shezidao can be traced back to 1970, when the Evaluation Report on Flood Control Schemes for the Taipei Area (WRPC-MEA 1970) was published. It was a study conducted by the Central Government, with some help from the United States Army Corps of Engineers, to determine the most economically and hydraulically feasible flood control scheme for the Taipei Area. The TAFCP was entirely based on the scheme proposed in the 1970 report, which excluded Shezidao from the 200-year flood protection. The report cites Shezidao's low economic output as a reason for its exclusion from the TAFCP (WRPC-MEA 1970, 89). This was because at the time Shezidao's economy was dominated by agriculture. In the 1970 report Shezidao was defined as a "flood zone"—a land area subject to flooding during extreme events—"and it shall remain agricultural land use and be restricted from any urban development" (WRPC-MEA 1970, 74). The definition of Shezidao as a flood zone resulted in its designation as a "development restricted zone" in Taipei City's overall master plan announced later in the same year of 1970.

Exclusion from the TAFCP has had an enduring impact on Shezidao. Since 1970, while the protected portion of the Taipei Area became increasingly urbanized and the living environment greatly improved, Shezidao has progressed little because its physical development has been legally prohibited. Today, Shezidao has often been described as being "frozen in time" (e.g., Yang 2016, 112).

The Unequal Protection within the Taipei Area

Because Shezidao is not included in the TAFCP, it is also dubbed "outside the levees," whereas the protected areas are "inside the levees" (Lin 2009). Such a depiction of Shezidao is nevertheless not entirely accurate. Shezidao has received some protection by FCI, but the protection standard has been far lower than that of the TAFCP. Shezidao residents have been mindful of such unequal protection and its implication in the imposed higher flood risk.

In 1975 the Taipei City Government built 2.5-meter-high tide embankments at selected locations in Shezidao to protect existing settlements with increasing populations against a five-year flood. Later, in 1978, these embankments were raised to 4 meters in response to the increased flood risk after the completion of the embankment of Zhumei, located to the east of Shezidao across the Keelung River (Figure 10.2). However, because such protection was much lower than that of the TAFCP, some of Shezidao's residents continuously pleaded for "inclusion in the TAFCP," that is, for FCI that can protect Shezidao against a 200-year flood. The request was repeatedly denied by the Central Government, citing the reason that it may "impact the TAFCP." Since Shezidao is a flood zone, protecting it against a 200-year flood would diminish the effect of the TAFCP by reducing the room for floodwater to increase the flood stage, thereby increasing the possibility of existing levees being overwhelmed.

In the 1980s, Shezidao's population continued to increase despite it being a development restricted zone. In 1986, it was found that the existing 4-meter-high embankments had subsided by an average of 30 centimeters (Taipei City Government 1987). Meanwhile, Shezidao residents grew increasingly agitated by the unequal protection. The anger finally exploded in March 1986 during the construction of a sluice gate—a TAFCP project—at the entrance of Shezidao (Yang 2016; Figure 10.2). The sluice gate, when closed during extreme events, would completely isolate Shezidao from the rest of the city. The construction of the sluice gate further reminded Shezidao residents of the fact that they are "outside the levees" and enhanced their fear of being left out during a large flood. The sluice gate was eventually not realized and instead a giant slope at the same location was built as an alternative. In 1987, the flood risk of Shezidao was further increased after the completion of the levees for Shanchong and Luzhou—also TAFCP projects—across the Danshui River in New Taipei City (Taipei City Government 1987; Figure 10.3). Some of Shezidao's residents staged several public protests to demand better FCI (Yang 2016).

In 1987, the Central Government finally authorized a capital project to encircle the entire Shezidao with a 6-meter-high levee system to protect it against a 20-year flood, as hydraulic modeling suggested that such a levee system would have little impact on the TAFCP (MEA 2010). Approval of the project came with a stipulation that Shezidao residents shall not request that the height of the levee system be raised further and that population growth in Shezidao should be restrained by land use regulations, because the protection standard is still low (MEA 2010). Construction of the 6-meter-high levee system started in 1989 and was completed in 1997. It has not been reinforced since. However, given that there is still a huge gap in the protection standard between Shezidao and the rest of the city, in the following years, the more vocal Shezidao residents continued to call for the inclusion of Shezidao in the TAFCP (Yang 2016).

Shezidao, after 52 Years of Development Restriction

As mentioned earlier, physical development in Shezidao has been legally prohibited following its designation as a flood zone and thus, a development restricted zone. The prohibited physical development includes the provision of public infrastructure, construction of new buildings, and retrofits of existing buildings. Moreover, the Taipei City Government used to require those who wanted to move to Shezidao to sign an affidavit, in which they would agree to bear responsibility should they suffer from any flood loss in Shezidao. While it is reasonable to avoid development in a flood-prone area so as not to put more people and property in harm's way, the development restriction since 1970 has created many difficulties for Shezidao residents because the existing living environment continues to deteriorate. Today, Shezidao is characterized by substandard living conditions, especially when compared to the rest of Taipei.

Dilapidated, Cramped, and Informal

Several factors have contributed to the substandard living environment in Shezidao. First, because building retrofit is prohibited, only "non-permanent" repair is allowed, and any repair that is considered "permanent" could lead to the entire building being declared as illegal. However, old buildings are prone to damage, particularly during the typhoon season. Over the past years, as residents could only apply makeshift repairs, numerous buildings erected before 1970, particularly those that have suffered severe structural damage, have deteriorated to an extent beyond habitable conditions.

Secondly, since no new building can be built, as families grew over time, more and more people have been forced to live in the same building, many of which were already small to start with. This has produced an unusual situation in Shezidao, referred to as "one address, multiple households." This applies to many core families – who would otherwise have lived as different households at different addresses when they branched off the original family – still living in the same building at one address. As of January 2019, there are 4282 households but only 1965 addresses in Shezidao; and out of the 1965 addresses, 296 are associated with more than three households.[1] It is not unusual to have as many as seven or eight households at one address (Author's personal communication with local residents). An extreme case is 15 households at one address.[2]

If one were to fully obey the development restriction, it would be almost impossible to live in Shezidao. Over the decades, many households have added additional structures on top of or adjacent to the existing buildings to create more room to house more family members. A few residents with ties to the local councilors and legislators managed to construct entire new buildings. Taipei City Government has chosen to close one eye to the illegal structures. Despite the government's tacit tolerance, most of the illegal structures are still makeshift in nature. This is because Shezidao's residents anticipate that their buildings will be demolished in the near future, since over the decades the Taipei City Government has repeatedly promised to redevelop Shezidao; this will be discussed later. In reality, the development restriction has not really frozen Shezidao in time, but instead, has led to numerous illegal, makeshift structures over the decades.

The "Unregulated Land"

Although as a flood zone Shezidao's land use should remain agricultural, in the early 1980s, manufacturing industries started to appear in Shezidao

1 Census data from the Department of Land Administration, Taipei City Government. https://www-ws.gov.taipei/Download.ashx?u=LzAwMS9VcGxvYWQvNDQ5L2NrZmlsZS83OTY3MzJlMiozZWUzLTQ5M2EtODA2My1kODUzYTAyNjcoNmQucGRm&n=MTA45bmoMeaciOS7vealtuexjee1seioiOizh%2baWmS5wZGY%3d&icon=.pdf, accessed 5 April 2019.

2 Census data from the Department of Land Administration, Taipei City Government. https://www-ws.gov.taipei/Download.ashx?u=LzAwMS9VcGxvYWQvNDQ5L2NrZmlsZS83OTY3MzJlMiozZWUzLTQ5M2EtODA2My1kODUzYTAyNjcoNmQucGRm&n=MTA45bmoMeaciOS7vealtuexjee1seioiOizh%2baWmS5wZGY%3d&icon=.pdf, accessed 5 April 2019.

(Wang 2001). By the mid-1990s, on many farmlands, large, cheap tin buildings for manufacturing had been built despite such land use being illegal in Shezidao (Hsieh 2008). This was against a backdrop where, on the one hand, Shezidao's agriculture was becoming increasingly uncompetitive compared to that in southern Taiwan, after completion of the first highway linking Taipei in the north to southern Taiwan; and on the other hand, Taipei was moving towards more strict land use regulations to remove factories from residential areas to increase the quality of life, despite Taiwan as a whole continuing to promote industrial development for economic growth (Hsieh 2008). Shezidao attracted many of the relocated factories, as well as new factories, all of which took advantage of Shezidao's relatively cheap land value and rent as a result of it being a development restricted zone (Hsieh 2008). Another factor was land-owning farmers in Shezidao being eager to find another livelihood, constructed large tin buildings, rented them out for manufacturing, which provided them a much more lucrative income (Hsieh 2008).

Today, there are 286 factories, only four of which are legal (Taipei City Government 2018). In the earlier years, Shezidao's factories were dominated by paper making, metalworking, and printing industries that have allegedly caused Shezidao's waters and soil to become heavily polluted (Taipei City Government 2018). No cleaning up has ever been attempted. In recent years, as the aforementioned industries declined, furniture making, warehousing, logistics, and recycling have become more popular (Taipei City Government 2018). This has resulted in numerous heavy trucks coming in and out of Shezidao, making its main street—the 6-meter-wide Yanping North Road (Figure 10.2)—and other even narrower streets dangerous, creating a constantly dusty and noisy environment for the street-front residents. Meanwhile, Shezidao also became a convenient—albeit illegal—dump site for construction debris. This resulted in the appearance of numerous mounds consisting of dubious materials in Shezidao's otherwise flat terrain.

The illegal factories and dumps have led Shezidao to be labeled as "unregulated land" (Yang 2016). It would be impossible for any of the factories and dumps to appear in a flood zone without the government's lax execution of the development restriction in Shezidao.

The Stereotypes of Shezidao

Under many unfavorable living conditions, those who are economically more capable have long moved out of Shezidao, leaving behind those either too old or too poor to leave (Author's personal communication with local residents). This also resulted in many unattended buildings rapidly deteriorating. Meanwhile, in the 1980s and 1990s Shezidao also became home to many migrant workers, who came to Taipei from the more rural parts of Taiwan in search of better incomes (Wang 2001). Like the illegal factories, they were drawn by Shezidao's much cheaper rent, compared to other parts of Taipei. Currently, many socioeconomically disadvantaged people and minorities reside in Shezidao (Lin 2009; Yang 2016). For the past 20 years Shezidao has had a stable population of around 11,000 (Taipei City Government 2018). However, it is believed that the number of people actually living in Shezidao is much smaller (Author's personal communication with local residents).

As the rest of Taipei City and other parts of Taipei Area continue to become urbanized, with the quality of life greatly improved, to outsiders Shezidao appears rural, informal, and even chaotic. Because of such contrast between Shezidao and the rest of the city, Shezidao is often depicted in the media as "dilapidated" or "backward."[3] Over the years, growing with Shezidao's deteriorating environment are a sense of inferiority and the stigma that people from Shezidao are poor. Shezidao residents often ridicule themselves as "second-class citizens" because of the government's differential treatment (Author's personal communication with local residents).

Developing Shezidao as a Solution?

Because of the distinct gap in the quality of the living environment between Shezidao and the rest of the city, "developing Shezidao" has been considered an ultimate solution to Shezidao's plight, even though Shezidao is a flood zone in the TAFCP and is supposed to be restricted from urban development. In the early years, when land development was an engine for economic growth and urbanization was considered a major way to improve the quality of life, it was generally accepted by the public that a rural area is eventually to be developed into an urban area. Today, many in Taiwan still equate urban development with a better standard of living. Although Shezidao is far from an undeveloped area, the Taipei City Government and the media

3 There have been numerous news reports, magazine articles, TV reports, and documentaries on Shezidao and the problem of development restriction. The description of Shezidao as "dilapidated" (殘破/頹敗/荒廢) can be seen, for example, in a television program aired on 19 January 2015 in TVBS News (https://news.tvbs.com.tw/politics/563584) and a news article in the Storm Media on 18 April 2015 (https://www.storm.mg/article/46349). The description of Shezidao as "backward" (落後) can be seen in a news article in NOWnews on 27 September 2018 (https://www.nownews.com/news/20180927/2984307/). Links accessed 5 April 2019.

continue to call it "the last virgin land in Taipei City";[4] hence the widely used expression of "developing Shezidao," although "redeveloping" would be more apt.

The Flood Control Controversy

Having had to endure the substandard living environment, Shezidao's local residents have been longing for the redevelopment of Shezidao to solve their plight. Since 1988, there have been six mayors of Taipei City, all of whom proposed urban development plans for Shezidao. However, none of these plans have materialized so far. The implementation of an urban development plan in Taiwan requires the approval and the subsequent official announcement of not only a master plan, but also a detail plan, which is often formulated after the announcement of the master plan. The detail plan also needs to be approved by the local government and then announced before the urban development plan can be carried out. In fact, Shezidao's very first master plan was announced in 1970, when Shezidao was also designated as a '"development restricted zone" in Taipei City's overall master plan. The 1970 master plan of Shezidao was dominated by agricultural and residential land uses. The detail plan was not formulated because the master plan stipulated that a flood control scheme for Shezidao must be approved by the Central Government before the detail plan can be approved. However, until Shezidao's flood protection is no longer considered a threat to the TAFCP, it will not be possible for any urban development plan to be materialized.

After the 6-meter-high levee system—which currently protects Shezidao—was approved by the Central Government in 1987, the Taipei City Government deemed that some urban development can take place in Shezidao (Taipei City Government 2010). In 1988, it proposed to develop Shezidao into a red-light and gambling district, because Shezidao's marginal and relatively isolated location makes it easier to concentrate and control these businesses. This development plan met strong resistance from local residents and women's rights organizations (Yang 2016). The master plan was eventually disapproved. In 1993, the Taipei City Government announced a different master plan with low-density development because Shezidao was only under low-level protection. However, because

4 A Google search using "Shezidao" (社子島) and "virgin land" (處女地) renders 95,300 results. Most of the entries are news reports related to the redevelopment of and real estate investment in Shezidao.

low-density development implies low buildable floor areas and thus lower gains, Shezidao's landowners, in particular, were dissatisfied with the announced plan (Chi 2011). In 1996, after two residents' assemblies, a conclusion was reached to appeal to the Taipei City Government to suspend the statutory procedure of announcing the detail plan; subsequently, in 1998, the Taipei City Government officially retracted the development plan (Taipei City Government 2018).

If Shezidao were to see high-density development similar to elsewhere in Taipei City, the Central Government would require high-level protection of Shezidao against a 200-year flood—like the rest of the city—to ensure flood safety. However, as mentioned earlier, the Central Government was concerned that the high-level protection of Shezidao might impact the TAFCP. To clarify this flood control controversy, the Taipei City Government commissioned a hydraulic-modeling study in 1998 to assess the potential impacts of the high-level protection of Shezidao. The study concluded that high-level protection of Shezidao against a 200-year flood would require the existing levee system to be raised to 9.65 meters, and in order for such high-level protection to make little impact on the TAFCP, the following five conditions would need to be met to maintain the flood conveyance capacities of the Keelung and Danshui Rivers: (1) the levee in Shezidao along the Keelung River would need to be set back by 80–130 meters; (2) the levee in Shezidao along the Danshui River would need to be set back by 30 meters; (3) the decommissioned landfill in Luzhou along the Danshui River would need to be removed; (4) the decommissioned landfill in Shanchong along the Danshui River would need to be removed; (5) and Wugu would need to be protected against a 200-year flood (Figure 10.4).

For the Taipei City Government, the hydraulic-modeling study pointed to an important way out of the flood control controversy that impeded Shezidao's urban development for the past decades. Subsequently, based on the study results, it proposed a different master plan with high-density development, in conjunction with a flood control scheme to protect Shezidao against a 200-year flood. In 2002, the master plan was approved, but it was not announced until nine years later, in 2011, after the Central Government in 2010 finally approved the flood control scheme for Shezidao upon which the announcement of the master plan was contingent. Besides the 9.65-meter levee system, the flood control scheme also involved raising the ground elevation of almost the whole of Shezidao to 8.15 meters above sea level, which would require 16.25 million cubic meters of soil. Because of concern regarding the potentially high environmental impact of such large-scale earthwork, the flood control scheme was eventually rejected

Figure 10.4 Map showing five conditions for Shezidao not to impact the TAFCP (Map adapted by Yin-Ling Huang)

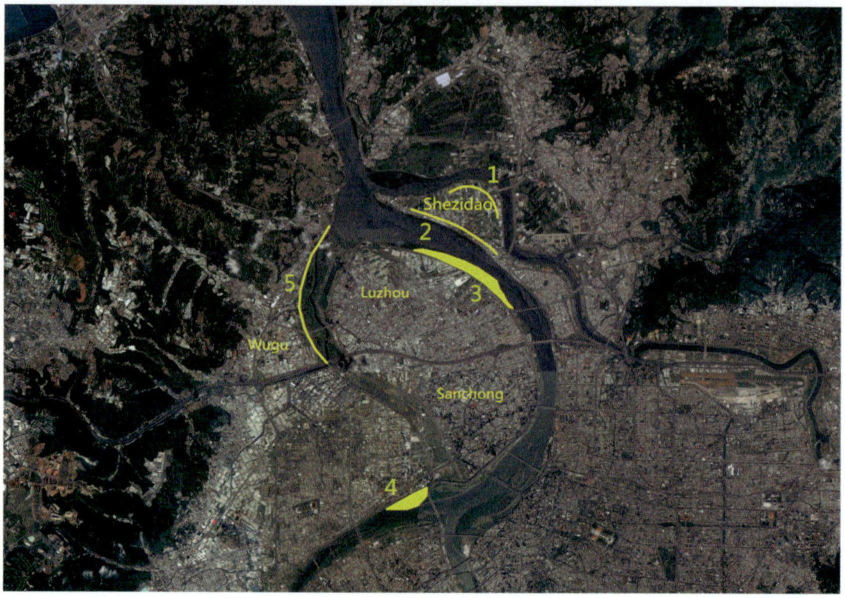

A study in 1998 concluded that in order for the 9.65m levee system not to impact the TAFCP, five conditions have to be met (1) the levee in Shezidao along the Keelung River needs to be set back by 80-130m; (2) the levee in Shezidao along the Danshui River needs to be set back by 30m; (3) the decommissioned landfill in Luzhou along the Danshui River needs to be removed; (4) the decommissioned landfill in Shanchong along the Danshui River needs to be removed; (5) and Wugu needs to be protected against a 200-year flood.

in the environmental impact assessment. Shezidao's redevelopment was again suspended.

The Current Redevelopment Plan

In 2015, the sitting mayor of Taipei City proposed yet another master plan, with a spatial plan completely different to the previous one. The flood control scheme for Shezidao was also revised. While it still includes the 9.65-meter-high levee system, the amount of land filling is to be significantly reduced to address the aforementioned concern regarding the earthworks. This latest proposed master plan was approved by the Central Government on 26 June 2018. Its detail plan was approved on 26 March 2020.

However, the master plan—hereafter referred to as the 2018 master plan—is highly controversial mainly because it involves the "zone expropriation" of the whole of Shezidao. Zone expropriation is a unique policy tool of land resumption in Taiwan, which allows the government to resume a contiguous area of land that involves multiple landowners. The 2018 master plan involves displacing all approximately 11,000 Shezidao residents, as well as removing all of the existing houses, factories, temples and shrines, and farmlands, except for five historical buildings and two larger temples (Taipei City Government 2018, 47). Such a redevelopment plan implies the complete eradication of the settlements that have existed for more than a century, along with the intangible community network and cultural heritage.

The Taipei City Government claims that all Shezidao residents will be resettled within Shezidao when the redevelopment is completed; however, the current compensation/resettlement scheme worries many residents. Importantly, it is not clear how the aforementioned situation of "one address, multiple households" can be appropriately addressed. The convention of compensation/resettlement associated with a redevelopment plan is that only one family at a particular address is entitled to any form of compensation/resettlement. This implies the displacement of many families in Shezidao. Moreover, most of Shezidao's residents—despite having lived in Shezidao for their entire lives—do not own the land. Many of them own only the building but not the land; others are long-term tenants. Many are lower- and low-income people, who might no longer be able to afford the much higher housing and rental prices after the redevelopment.

While the 2018 master plan could improve the deteriorating living environment of Shezidao, it is unofficially estimated that it might result in one third of the residents being driven out of their homes (Author's personal communication with local residents). Ironically, if the 2018 master plan were to be materialized, many of those who have been suffering from the deteriorating living environment might be unable to enjoy the fruits of redevelopment.

The past decades have seen the Taipei City Government's repeated attempts to solve Shezidao's problem of the deteriorating environment by "developing Shezidao." Decades have gone by, and today Shezidao residents are still living with the development restriction imposed since 1970. The repeated failures of the government to materialize the redevelopment, coupled with the fact that promises of Shezidao's redevelopment were often made during mayoral elections, have resulted in the loss of Shezidao residents' trust in the government. Many no longer believe that the development of Shezidao will ever happen. As Shezidao residents continue to endure the deteriorating environment, they have developed a strong sentiment of abandonment and injustice based on the belief they are being sacrificed for the flood safety of the Taipei Area (Author's personal communication with local residents).

Discussion

The decision of where FCI should be built to protect whom involves value judgment, and is not just a technical exercise (Chan and Liao 2022). The decision made in 1970 by the Central Government to exclude Shezidao from the TAFCP was based solely on economics through the use of a cost-benefit analysis. Such decision making has been considered rational and is rarely challenged. Regardless of whether FCI is in fact effective for reducing flood risk, we must ask: should the protection of life be based solely on its economic worth? Are the lives of people who live in rural areas—such as those in Shezidao—less worthy of protection?

However, the fact that Shezidao is not protected by FCI against a 200-year flood is not the issue itself. The issue raised in this chapter is floodwater redistribution and consequential flood risk transfer from the economically more valuable areas to Shezidao through the government's differential treatment with its flood control plan. Because of the unequal protection within the Taipei Area, Shezidao would be forced to bear more floodwater during a 200-year flood, while the rest of the Taipei Area remains free of floodwater. This leads to environmental injustice (Liao et al. 2019). Consider a "natural" condition of the Taipei Area: most of the lowland area is subjected to the periodic flooding of the Danshui and Keelung Rivers and no FCI exists anywhere; when the river floods, the floodwater would be more evenly distributed across the Taipei Area. Nevertheless, today, the much higher levees elsewhere in the Taipei Area would transfer floodwater and the associated flood risk to Shezidao during any flood that is beyond Shezidao's flood control capacity. Somehow, this flood risk transfer was inexplicitly justified by the cost-benefit analysis.

As discussed earlier, Shezidao's redevelopment was affected by a flood control controversy, which was that Shezidao's development might impact the TAFCP. However, taking into account the history of the TAFCP, a more precise discourse should instead be: if Shezidao stopped bearing the extra, transferred floodwater, the levees elsewhere in the Taipei Area could be overwhelmed. To deny Shezidao high-level protection against a 200-year flood implies that Shezidao has to continue to act as a flood retention area for the rest of the Taipei Area. There is no data or modeling result available to show how much extra floodwater the TAFCP would redistribute to Shezidao during a flood of a given magnitude. It is unlikely to be a small amount during a 200-year flood, judging by the results of the aforementioned hydraulic study in 1998. Based on the study, if Shezidao ceased to be a flood zone, a lot of extra space would need to be created in both the Danshui and Keelung Rivers to ensure the efficacy of the TAFCP (Figure 10.4). The extra space therefore indicates the amount of floodwater the TAFCP would redistribute to Shezidao, if a 200-year flood were to occur.

Shezidao residents have been well aware of the fact that Shezidao would act as a flood retention area for the rest of the Taipei Area during a large flood. As tax-paying citizens, they have felt disadvantaged by the government's preferential treatment, hence the self-ridicule of second-class citizens. Nevertheless, developing Shezidao and protecting it with better FCI appear not to be a solution to Shezidao's plight. While the 2018 master plan could protect Shezidao with the same level of protection as the rest of the city, it would not bring the justice due to Shezidao residents but instead displace many of those who have suffered as second-class citizens for more than half a century. This is because zone expropriation of the entire Shezidao is proposed as the only development tool, the reason for which, ironically, is to finance the land resumption needed for the 9.65-meter-high levee system, long called for by some of Shezidao's residents. Many Shezidao residents, therefore, would not be able to enjoy the protection of this levee system because they would be driven out of Shezidao by the redevelopment. Even more ironically, the Taipei City Government claims that building this levee system will provide "social equity and justice" for Shezidao (Taipei City Government 2010, IV).

Although the marginalization of Shezidao is largely due to the unequal protection within the Taipei Area, making the protection equal by upgrading FCI cannot be a solution. Besides the problem of displacement, the proposed 9.65-meter-high levee system for Shezidao would cause further floodwater redistribution within the watershed. Even if all the aforementioned five conditions were met (Figure 10.4) and the 9.65-meter-high levee system for Shezidao ended up not impacting the TAFCP, the current flood protection standard for the Taipei Area would eventually become inadequate in the face of climate change. Since it is impossible to reinforce the FCI everywhere at once, the problem of who should be protected and at whose expense will remain. Socioeconomically marginalized communities, such as Shezidao, will still be the first to be sacrificed.

Unfortunately, the case of Shezidao is far from unique. Around the world similar cases can be seen. For example, flood control for Bangkok's central business district—where the majority of the city's higher-income people live and work—has indirectly rendered the suburban neighborhoods and the upstream farmlands retention areas for excess floodwater (Lebel and Sinh 2009). Such differential treatment is more conspicuous during emergency events. During the 2011 catastrophic flood of central Thailand, the Thai government strived to keep the inner part of Bangkok and certain industrial estates dry, leaving its western and northern outskirts at similar elevations heavily flooded for many weeks (Marks 2015). In Vietnam, it is a policy that during an emergency, excess floodwater in the Red River would be diverted from the capital city of Hanoi to nearby provinces, despite this impacting as many as 675,000 people (Lebel and Sinh 2009). Elsewhere, in the West, during the Great Mississippi Flood of 1927, in order to save New Orleans, levees downstream of the city were artificially breached and millions of gallons of river discharge entered the wetlands to allow the flow to pass through the city quickly; however, this drowned the commercially important muskrats and disrupted oyster cultivation in the wetland (Colten 2007). Similarly, in 2011, the U.S. Army Corps of Engineers blasted a levee in Missouri to divert floodwater in the Mississippi River away from Cairo, Illinois—a town of 3000 people, leading to severe damage to 130,000 acres of farmland and approximately 100 million US dollars in crop losses (Eversley 2011). In that same year, the Manitoba Government of Canada breached a dike along the Assiniboine River and damaged farmlands in order to reduce flooding downstream (CBC 2011). The ideology that socioeconomically less important areas should be sacrificed for socioeconomically more important ones is shared across the world in modern society.

Conclusion

Acting as a flood zone in the TAFCP, residents in Shezidao have endured a substandard living environment and the accompanying stigma for half a century, while the rest of Taipei City enjoys the fruits of the TAFCP. The case of Shezidao demonstrates how flood control can benefit some places while disadvantaging others through floodwater redistribution and flood risk transfer. This social downside of flood control affects many other socioeconomically marginalized communities around the world. This needs not only to be acknowledged, but also to be addressed (Chan and Liao 2022).

What, then, would be a plausible solution to Shezidao's plight? It should be pointed out first that the Taipei City Government's failure to resolve Shezidao's problem of deteriorating environment is mainly because it has been fixated on two notions: urban development is the only solution; and flood control is necessary for urban development. Both are fallacies. First, it is possible to improve Shezidao's deteriorating environment through retrofits, instead of the total erasure of its existing environment by starting anew. Secondly, and more importantly, flood control is hardly the only option for flood safety. Rarely do people challenge the norm of flood control because flooding has largely been considered only a hazard. However, research suggests that flood exposure does not always lead to flood vulnerability (Walker and Burningham 2011). This was historically true in Shezidao. Before its existing levee system was built, floods rarely caused damage, even though Shezidao was frequently flooded. This was because the earlier settlers in Shezidao understood the risk arising from living in a flood-prone area and adapted their lifestyle to frequent flooding (Author's personal communication with local residents). Today, there are still vernacular settlements in other parts of the world that live peacefully with periodic flooding by adapting to, instead of controlling it (Laituri 2000). For example, many rural villages in the Mekong Delta not only use stilt houses, boats, temporary footbridges, etc., to remain functional during prolonged seasonal flooding, but also have developed livelihoods dependent on flooding (Liao et al. 2016). Because flood exposure does not necessary lead to danger, flood control is hardly the only solution to flood safety.

In recent years, nonstructural measures have received increasing attention in flood management. Nonstructural measures—particularly flood warning systems, better evacuation and recovery planning, and flood insurance—should be implemented to enhance Shezidao's flood resilience in the face of climate change. Furthermore, flood adaptation can be an alternative to flood control (Liao 2014). Instead of increasing the height of the existing levee system in Shezidao, the Taipei City Government could help retrofit Shezidao's existing environment and adapt it to riverine flooding. The adaptive measures include elevating the buildings or flood-proofing the lower floors of the buildings, designing open spaces that also function as flood retention basins, and designing a transportation system and roadways that can operate in dry and flooded conditions (Liao 2014). In river cities, riverine flooding can never be eliminated, and climate change has already made frequent flooding a norm rather than an exception. A paradigm shift from flood control to flood adaptation not only is a realistic approach to climate change adaptation, but can also fundamentally address the problem of floodwater redistribution and flood risk transfer. Instead of relying on the centralized mechanism of FCI to prevent flooding for selected communities, every community in a naturally flood-prone area should be "responsible for its own floodwater." That is, each community—and each citizen—should try to accommodate locally the floodwater that naturally occurs.

The paradigm of flood adaptation, which stresses local liability, might sound radical, but it is not. Local liability has been the essence of the recent paradigm shift in urban storm water management, which has been manifested as Low Impact Development in the US, Water Sensitive Urban Design in Australia, and Sustainable Urban Drainage System in the UK, among others. In contrast to the conventional paradigm that focuses on efficient drainage but simply shifts the problems of storm water runoff elsewhere, this new paradigm focuses on tackling storm water runoff at the source to improve its quality and to avoid polluting the receiving water and to reduce its quantity to avoid downstream flooding (Liao et al. 2017). Many cities and nations around the world now have a drainage code that requires new development to retain the increased storm water runoff on site. This is to ask each community to be responsible for its own storm water runoff such that it does not become others' problem. As the liability associated with storm water runoff—also a source of flood hazard—has been increasingly recognized, attention should also be paid to the liability associated with riverine floodwater. Citizens of river cities should acknowledge the fact that the river floods naturally and should try to handle the "given" floodwater locally, instead of using FCI to redistribute floodwater and harm others.

Admittedly, in today's context it is extremely difficult to ask individual urban communities to be responsible for their own floodwater, since FCI has long been established in urban centers around the world, and those who have been protected by FCI have become too used to it. Nevertheless, acknowledging the problem of floodwater redistribution and flood risk transfer is the first step toward a more just world, where the socioeconomically marginalized would no longer have to endure the downside of flood control.

Acknowledgments

I thank Yin-Ling Huang for producing all the figures in this chapter and Guan-Tsen Lin for helping collect the archives of government plans and reports pertaining to the TAFCP. I am also indebted to the residents in Shezidao, whom I interviewed, and to many others whom I have encountered in my observational study. They helped me tremendously to better understand Shezidao. Lastly, I thank Jeffrey Kok Hui Chan from the Singapore University of Technology and Design, as well as the editors and reviewers of this volume, for providing valuable comments to help improve this chapter.

References

Brookes, Andrew. 1988. *Channelized Rivers: Perspectives for Environmental Management*. New York, NY: Wiley.

CBC. 2011. "Water Flow through Manitoba Dike Breach." *CBC*, 15 May. http://www.cbc.ca/news/canada/manitoba/water-flows-through-manitoba-dike-breach-1.975820, accessed 30 March 2019.

Chan, Jeffrey Kok Hui, Liao, Kuei-Hsien. 2022. "The normative dimensions of flood risk management: Two types of flood harm." *Journal of Flood Risk Management*, 15(2): e12798.

Chi, Ping. 2011. *The Study of Stakeholders' Attitude to the Development of Shezidao* (in Chinese). Master's thesis. Department of Information Management, National Sun Yat-sen University.

Colten, Craig E. 2007. "Environmental Justice in a Landscape of Tragedy." *Technology in Society*, 29: 173–179.

Douglas, Ian, Kurshid Alam, Maryanne Maghenda, Yasmin McDonnell, Louise McLean, and Jack Campbell. 2009. "Unjust Waters: Climate Change, Flooding and the Urban Poor in Africa." In *Adapting Cities to Climate Change: Understanding and Addressing the Development Challenges*, edited by J. Bicknell, D. Dodman, and D. Satterthwaite, 201–223. London: Earthscan.

Eversley, Melanie. 2011. "Mo. Levee Blast Inundates Acres of Farmland." *USA Today*, 4 May. http://usatoday30.usatoday.com/weather/floods/2011-05-03-mississippi-river-cairo-illinois-levee-floods_n.htm, accessed 30 March 2019.

Hewitt, Kenneth, and Ian Burton. 1971. *The Hazardousness of a Place: A Regional Ecology of Damaging Events*. Toronto: University of Toronto Press.

Hsieh, Mei-Hua. 2008. *Urban Regime, Planning Politics, and the Emergence of Citizen Consciousness: The Imagination and Reality of Development for Shezidao* (in Chinese). Master's thesis. Graduate Institute for Social Transformation Studies, Shih Hsin University, Taiwan.

Laituri, Melinda. 2000. "Cultural Perspectives of Flooding." In *Inland Flood Hazards: Human, Riparian, and Aquatic Communities*, edited by E.E. Wohl, 451–468. Cambridge: Cambridge University Press.

Lebel, Louis, and Bach Tan Sinh. 2009. "Risk Reduction or Redistribution? Flood Management in the Mekong Region." *Asian Journal of Environment and Disaster Management*, 1(1): 25–41. http://doi.org/10.3850/S179392402009000040.

Liao, Kuei-Hsien. 2014. "From Flood Control to Flood Adaptation: A Case Study on the Lower Green River Valley and the City of Kent in King County, Washington." *Natural Hazards*, 71: 723–750.

Liao, Kuei-Hsien, Jeffrey Kok Hui Chan, and Yin-Ling Huang. 2019. "Environmental Justice and Flood Prevention: The Moral Cost of Floodwater Redistribution." *Landscape and Urban Planning*, 189: 36–45.

Liao, Kuei-Hsien, Shinuo Deng, and Puay Yok Tan. 2017. "Blue-Green Infrastructure: New Frontier for Sustainable Urban Stormwater Management." In *Greening Cities: Forms and Functions*, edited by P.Y. Tan and C.Y. Jim, 203–226. Singapore: Springer Nature Singapore.

Liao, Kuei-Hsien, Anh Le Tuan, and Kien Van Nguyen. 2016. "Urban Design Principles for Flood Resilience: Learning from the Ecological Wisdom of Living with Floods in the Vietnamese Mekong Delta." *Landscape and Urban Planning*, 155: 69–78.

Lin, Su-Wen. 2009. "Shezidao: Neighbourhood, Community, or Local Society?" (in Chinese). The 26th Seminar of the School of Liberal Arts and Social Sciences, Soochow University.

Marks, Danny. 2015. "The Urban Political Ecology of the 2011 Floods in Bangkok: The Creation of Uneven Vulnerabilities." *Public Affairs*, 88(3): 623–651.

Mata, Rodolfo. 1993. "Inequitable Siting of Undesirable Facilities and the Myth of Equal Protection." *Boston College Third World Law Journal*, 13: 233–256.

Ministry of Economic Affairs (MEA). 2010. *Recommended Flood Control Scheme for Taipei Area* (in Chinese). Taipei City: Taiwanese Ministry of Economic Affairs.

Padawangi, Rita, and Mike Douglass. 2015. "Water, Water Everywhere: Toward Participatory Solutions to Chronic Urban Flooding in Jakarta." *Public Affairs*, 88(3): 517–550.

Pinter, Nicholas, Fredrik Huthoff, Jennifer Dierauer, Jonathan W.F. Remo, and Amanda Damptz. 2016. "Modeling Residual Flood Risk behind Levees, Upper Mississippi River, USA." *Environmental Science & Policy*, 58: 131–140.

Postel, Sandra, and Brian Richter. 2007. *Rivers for Life: Managing Water for People and Nature*. Washington, DC: Island Press.

Taipei City Government. 1987. *The Levee Protection Plan for Shezidao* (in Chinese). Taipei City: Taipei City Government.

Taipei City Government. 2010. *An Analysis of the Benefits of the Development of Shezidao and Its Impact on the Taipei Area Flood Control Plan* (in Chinese). Taipei City: Taipei City Government.

Taipei City Government. 2018. *Rezoning of the Master Plan of Shezidao, Shilin District, Taipei City* (in Chinese). Taipei City: Taipei City Government.

Taylor, Dorceta E. 2014. *Toxic Communities: Environmental Racism, Industrial Pollution, and Residential Mobility*. New York, NY: New York University Press.

Tilt, Bryan, Yvonne Braun, and Daming He. 2009. "Social Impacts of Large Dam Projects: A Comparison of International Case Studies and Implications for Best Practice." *Journal of Environmental Management*, 90: S249–S257.

Tobin, Graham A. 1995. “The Levee Love Affair: A Stormy Relationship.” *Water Resources Bulletin*, 31(3): 359–367.

van Voorst, Roanne and Jörgen Hellman. 2015. “One Risk Replaces Another: Floods, Evictions and Policies on Jakarta’s Riverbanks.” *Asian Journal of Social Science*, 43: 786–810.

Walker, Gordon and Kate Burningham. 2011. “Flood Risk, Vulnerability and Environment Justice: Evidence and Evaluation of Inequality in a UK Context.” *Critical Social Policy*, 31: 216–240.

Wang, Chih-Wen. 2001. *On the Great Changes of Shezi: Community Empowerment*. Taipei City: Shezi Culture and Education Foundation.

WRPC-MEA: Water Resources Planning Committee. 1970. *Evaluation Report on Flood Control Scheme for Taipei Area* (in Chinese). Taipei City: Taiwanese Ministry of Economic Affairs.

Yang, Li-Ling. 2016. *Shezidao: A Forgotten Island, Prisoned for Half a Century*. New Taipei City: Walkers Cultural.

Biography

Kuei-Hsien Liao, Ph.D. is Associate Professor at the Graduate Institute of Urban Planning, National Taipei University, Taiwan. Her research focuses on the interaction between cities and rivers with respects to flood hazard mitigation, flood resilience, ecosystem services, and environmental justice.

11 Hybrid Riverscapes

Governing River Cities beyond Land and Water Dichotomies—the Yamuna in Delhi, India

Alexander Follmann

Abstract

The inherent complexity of environmental change in river cities requires a shift of perspective going beyond binary conceptualizations of water/land, river/city, and nature/culture. By linking a discourse analytical approach with theoretical concepts from governance research and urban political ecology, the framework of hybrid riverscapes—spatially referring to the riverine landscape formed by the natural forces of the river and human interventions—is outlined as an innovative concept to study environmental change and governance in river cities. Using the case study of Delhi's riverscapes and the construction of a large bus depot on the banks of the Yamuna, the chapter outlines how such a concept of riverscapes can advance fluid, liminal, and hybrid theory making on river cities.

Keywords: governance, urban political ecology, environmental change, discourse analysis

Introduction

For long, urban rivers have been viewed as elements of the urban landscape from a city-centric point of view. In doing so, the river-city nexus has been characterized around the physical space of the riverfront—the space of the city at the edge of the water marking a seemingly clear boundary between the city and the river. Yet, studies on the socionature of (urban) rivers have shown that river-city relationships are complex, and a "natural" or "fixed" boundary separating the river from the city consequently does not exist

Padawangi, Rita, Paul Rabé and Adrian Perkasa, *River Cities in Asia*.
Taylor & Francis Group, 2022
DOI: 10.5117/9789463721851_CH11

(Desfor et al. 2011; Follmann 2016a; Holifield and Schuelke 2015; Lahiri-Dutt 2014; Lahiri-Dutt and Samanta 2013; Rademacher 2011). The concept of socionature is used to indicate that nature and humanity are not divided but rather are ontologically inextricable. Additionally, "the production process of socionature," as Swyngedouw (1999, 447) highlights, "embodies both material processes and the proliferating discursive and symbolic representations of nature." Therefore, the inherent complexity of environmental change in river cities requires a change of perspective going beyond binary conceptualizations of water/land, river/city, and nature/culture.

By linking a discourse analytical approach with theoretical concepts from governance research and urban political ecology (UPE), this chapter outlines the framework of hybrid riverscapes (see also Follmann 2016a)—spatially referring to the riverine landscape formed by the natural forces of the river and human interventions—and develops an innovative conceptual framework to study environmental change and governance in river cities. Hybridity is a key frame in understanding river cities. As indicated in the introductory chapter, river cities refer to cities in which the river system—including creeks, canals, lakes, and wetlands—is an integral part of the city in all its historical, social, cultural, and political dimensions. As such, governing river cities as hybrids must go beyond land and water dichotomies. The central question hence is: How are hybrid riverscapes governed in river cities? UPE scholars have used a wide range of theoretical thoughts to reconceptualize the nature-society relationship to (re)politicize debates on urban nature and its uneven geographies. Based on the production of nature thesis by Neil Smith (1984), Latour's conceptualization of human and nonhuman actants and socionature (1993, 2004), and the hybrid geographies of Sarah Whatmore (2002), riverscapes are conceptualized as hybrids. In particular, Smith's production of nature thesis is used to challenge, rethink, and (re)politicize the contemporary (re)making of urban rivers across the world (Follmann 2016a; Loftus 2012). The detailed conceptualization of the riverscapes as hybrids draws on Erik Swyngedouw's (1999, 2004) concept of waterscapes connecting the material hydrological cycle of water with the (nonmaterial) landscapes of power constituting together the hydrosocial cycle of water (cf. Baviskar 2007; Karpouzoglou and Vij 2017). Thus, while existing research has often used the terminology of the waterscape, I prefer the term riverscape, "since it underlines the agency of the river in scaping its territory within and outside the city. The suffix scape, when understood as socio-nature, highlights that this creation is a result of a process of both human practices and the actions of the river (nature) itself" (Follmann 2016a, 83).

Existing studies on rivers in India's cities have long thematized issues of pollution (Alley 2012; Haberman 2006) or large-scale dams (D'Souza 2006; Baghel 2014). More recent studies on rivers in Indian cities have focused on urban floods (Arabindoo 2016; Bremner 2020) and large-scale riverfront projects (Coelho and Raman 2010, 2013; Desai 2012; Follmann 2016a; Pessina 2019). Outlining the contradictions of modernity's water-land binary, this chapter aims to add to this literature by using the case of a large bus depot constructed on the banks of the Yamuna in Delhi. Based on UPE's fundamental concept of the "politicized environment" (Bryant and Bailey 1997) the next section outlines a politicized understanding of the governance of river cities. Drawing on the work of Latour (1993, 2004), the third section outlines modern practices of hybridization and purification. The fourth section introduces the case study of the bus depot on the banks of the Yamuna in Delhi. The fifth section discusses the case study before presenting a final conclusion.

Politicizing the Governance of Urban Rivers through UPE

Based on the central understanding of the "politicized environment" (Bryant and Bailey 1997), UPE scholarship has radically critiqued technocratic governance approaches and highlighted that urban environmental governance is not only about proper management but equally about power and politics. Such a critique of normative concepts of (good) governance is particularly relevant for cities (and their rivers) in the Global South, as the discussion there has often been coined by techno-managerial, often imported, and frequently colonially coined apolitical approaches to governing urban nature (Lawhon et al. 2014; Myers 2008). Indeed, the dominant understanding of governance—as "a software techno-fix" (Hirsch 2006, 190)—focuses on coordination and collaboration rather than the conflicting perceptions and interests of the multiplicity of actors involved in urban environmental governance. In particular for cities in the South, it is well-documented that techno-managerial issues tend to dominate the governance of urban nature, thereby marginalizing "other" conceptualizations of nature in the city, especially those of the poor and marginalized groups (Karpouzoglou and Zimmer 2016; Keeley and Scoones 2003). While holding on to the term governance, governance is conceptualized here as a research object. In doing so, the analysis goes beyond questions of (better) coordination and collaboration among the actors linked to normative categories of good/bad governance. Rather, the analysis focuses on the interaction, negotiation, and bargaining processes among the different actors (Follmann 2016a, 65).

Such an analytical perspective on governance aims to look into the actual existing processes of governance with the goal to better understand the interactions among the different actors.

With regard to the underlying motives, driving forces, and powers of the actors involved, governance research has emerged to be of limited explanatory value. Hybrid concepts offer here the opportunity to connect an analytical governance perspective to recent post-structuralist insights from UPE. While research in this vein has deeply engaged with water, studies focusing on the decision-making processes for, and implementation of, spatial policies—especially also its links to, and synchronization with, environmental policies—is largely missing in particular for river cities in the Global South. As outlined elsewhere in more detail (Follmann 2016a, 61–67), by focusing on both "consensus-finding processes" and "causes of tension and conflict" (Hohn and Neuer 2006, 293) between the multiplicity of actors involved in policy making on urban nature, "an analytical approach to governance is able to overcome the risk of reducing governance to an apolitical organizational problem" (Follmann 2016a, 61).

In this context, urban environmental governance needs to be understood as a distinct field of policy making which brings together multiple spheres of governance at the local level while simultaneously taking into account multiple scales. Only such an approach is able to set in context and reconnect the often sectoral fragmentation of river governance within and beyond urban areas (Baghel 2014; Hartmann and Spit 2014). Existing governance research has often remained focused on sectoral approaches, such as studying water governance. In doing so, governance research has often failed to holistically conceptualize the socio-environmental processes leading to urban environmental change. Research on urban environmental governance has thus often focused on certain partnerships in urban environmental management in terms of service provisions, e.g., water supply, waste management, sewerage, etc. (cf. Baud and Dhanalakshmi 2007; Maria 2008; Zérah 2009).

The fragmentation of governance setups and the operation in sectoral silos is both a cause and effect of sectoral approaches in governing the urban environment. An analytical perspective questions how and why this fragmentation in urban environmental policies occurs; and why more holistically oriented policies to govern urban environmental change—if formulated at all—often remain pipe dreams. In this context, Hartmann and Spit (2014, 791) recommend that "water management steps into the governance arena of spatial planning, and spatial planning needs to reconsider its notions of water issues." In particular, they argue for overcoming the "institutional divide between spatial planning and water management," which often follow different rationalities, have very different institutional capacities, draw on different sets of knowledge, and follow their own sectoral agendas (Hartmann and Spit 2014, 791).[1] For example, the debate on "space for rivers" for flood protection indicates that river governance is essentially a spatial challenge—both in absolute physical/geographical terms of space as well as relational to other uses (cf. Wiering and Immink 2006, 423). Reserving open space for flood water is especially crucial for Asia's monsoonal rivers (cf. Bremner 2020, 748). Therefore, a coordination of sectoral approaches in governing urban rivers is highly needed.

More coordination and more knowledge, however, does not automatically solve problems of (urban) river governance (Billé 2008). First, urban environmental change in river cities is complex and floodplains are contested spaces (Lübken 2012). Second, multi-sectoral river governance is a contested terrain in which (sectoral) actors are keen to protect their (at best undisturbed) spheres of governance trying to avoid interference by other spheres and actors. These contestations are "territorial struggles" which "entwine battles over natural resources with struggles over meaning, norms, knowledge, decision-making authority, and discourses" (Boelens et al. 2016, 8). Therefore, UPE-informed governance research studies these processes of (institutional) territorialization around urban rivers and aims to deconstruct the logics and rationalities of the involved actors in order to overcome the naturalization of the land/water dichotomy and deconstruct hegemonic beliefs (Boelens et al. 2016; Follmann 2016a). In order to outline the underlying understanding of hybridity within the concept of hybrid riverscapes, the next section outlines the processes of hybridization and purification drawing on Latour's (1993, 2004) work.

Hybridization and Purification—Thinking beyond Land and Water Dichotomies

Modernity, as argued by Latour (1993, 10), is designated by "two sets of entirely different practices": "hybridization" and "purification." Hybridization produces hybrids by mixing nature and culture, while the separation of nature and culture is produced through purification—an active process of attributing hybrids to the pole of either nature or culture. In other words,

1 In the Indian context, both water resource management and spatial planning rely heavily on technical and engineering knowledge (Alley 2012; Baghel 2014; Iyer 2015).

"hybrids are the result of hybridization through the metabolism of human practices (in Marxist terms, human labor) and 'natural' processes" (Follmann 2016a, 79). This means, the physical materiality of urban natures—including urban rivers—is the result of the urbanization of nature through the process of urban metabolization. However, generally through omnipresent processes of purification, urban nature is purified into the categories of human/nonhuman, culture/nature, city/river, and land/water. These omnipresent processes of purification remain, however, often unquestioned and the outcomes are clear-cut boundaries between the above outlined dichotomies. In doing so, dichotomies are naturalized, and they are never neutral but rather always politically charged.

Purification is ever-present and unquestioned as it allows for—as well as legitimizes—modernist ontology to draw boundaries between dichotomous categories. As Forsyth (2003, 89) argues, boundary making "allows the establishment of an order and framework from which to proceed." Hybrids in contrast, as Latour (2004, 24) insists, "have no clear boundaries, no well-defined essences, no sharp separation between their own hard kernel and their environment." Thus, any hybrid—which is not, or not yet purified—creates contradictions, tension, and conflicts for modernist, dichotomous understandings of the world (Follmann 2016a, 78–79). Such boundaries are essential in order to govern urban nature from a techno-managerial perspective. For example, land-use planning—in particular, zoning—is fundamentally based on processes of purification (Whatmore and Boucher 1993, 169): "Land-use planning formalizes the separation between nature and abstract space through the written codes of legal statute and professional conduct which impose a site-based, rather than system-based, narrative structure on its treatment of the environment."

Notably, purification is about simplifications (Scott 1998) in order to make space governable. Therefore, Murdoch (2006, 133) highlights that land-use planning is the technology for the "'taming' of space," in other words, territorialization as it defines the physical geographies of nature and society, the river and the city.

Obviously, the changing morphologies of rivers and their floodplains due to their inherent fluidity have always posed difficulties for spatial classifications whether with regard to land ownership (Blomley 2008) or land-use zoning (Haidvogl 2012; Hartmann 2013; Lübken 2012). This is because land-use planning envisions, and is based upon, neat boundaries between different land uses, and in particular between land and water. In the modernist, engineered understanding the space of urban rivers must be determined as either land or water and a clear boundary—the riverfront—emerges on the plan. This is contrary to the understanding of riverscapes as hybrids. Using the concept of riverscapes as hybrids, however, helps to understand processes of purification which underlie, structure, and legitimize the governance of river cities. Research thus needs to unveil and deconstruct how the omnipresent processes of purification discursively (through language, institutions, regulations, etc.) and materially (through dams, embankments, channels, etc.) divide the hybrid riverscapes into the spatially defined territories of land and water.

The following case study of the riverscapes of the Yamuna in Delhi—with special attention to the planning, implementation, and post-implementation discourse around the Millennium Bus Depot—aims to highlight how powerful actors employ the tools of purification—both discursively and materially—as well as the contested and politicized environment of urban rivers in river cites more generally.

Case Study

Situating the Case: Urban Environmental Change along the Yamuna in Delhi

The River Yamuna, which is considered to be a holy river in Hindu mythology, takes its source in the glaciers of the Himalayas and joins with the Ganges in the city of Allahabad. The Yamuna flows for 48 kilometers through India's capital city, Delhi. The river's floodplain forms the largest near-natural feature of the megacity. The river's natural flow shows great differences in seasonality; while in the dry season hardly any flow occurs, major floods are common between July and September. Defined as the River Zone (Zone O in the Master Plan of Delhi), the remaining floodplain within the city limits is about 97 square kilometers large. Piece by piece, parcels of land within the floodplain have been protected by embankments for power plants, fly-ash ponds, sports facilities, public memorials, train depots, and enclosed residential compounds. However, different plans for comprehensive channelization schemes, which have been proposed to reclaim land for Western-style, large-scale riverfront development, have failed to be implemented in the last decades. Besides technical and financial reasons, environmental activism aiming to protect the river and its floodplain has played a major role in "rescuing" the remaining floodplain. Three major discourses (Follmann 2016b) are evident: (1) questions of minimum water flow in the river, (2) pollution amendment and ecological restoration; and most recently, (3) the preservation of the floodplain and its protection against built-up developments.

As the governance of the Yamuna has always been characterized by sectoral approaches, these three concerns have seldom been discussed in combination. In the last two decades, different environmental NGOs have, however, challenged the sectoral fragmentation of river governance in Delhi by linking scales and sectoral aspects (for more details on the role on environmental NGOs see Follmann 2016a, 2016b). Nevertheless, the Yamuna in Delhi is certainly one of the most polluted river stretches in the world, due to the obstruction of its natural flow for drinking water provisions and the discharge of mostly untreated or partially treated wastewater as a result of the under-capacity and non-functionality of the entire city's sewage treatment system.

For decades, the river was the backyard of the city and slums used to dominate the banks of the river. The flood-prone, polluted, and marginalized spaces were settled upon informally by the poor. Yet, triggered by Public Interest Litigations (PILs) in the courts, between 2004 and 2006 major slums were demolished on ecological grounds and hundreds of thousands were evicted from the river's banks. The slums were viewed as encroachers on public land, polluting the river's water and destroying the nature of the river (Baviskar 2011; Bhan 2009; Ghertner 2011; Sharan 2016). The court ruled: "It is required to be noted that the policy ... for [slum] relocation would not apply to the river bed. It is not an encroachment on land. It is a water body and it is required to be maintained as a water body" (High Court of Delhi, 3 May 2005, WP (C) No. 689/2004). Therefore, in the discursive framing the "land" occupied by the slums along the river was purified by the court as "water body," indicating that the space should be given back to the river (Follmann 2016a, 235). The High Court of Delhi further set a 300-meter no-development zone along the banks of the river—demarcating the land of the river. However, in practice, the land-owning agencies—mainly government departments—only tackled the slums; other structures remained on the floodplain. Even large-scale developments were carried out on the floodplain in the 2000s, including new Metro train depots and the sports and residential facilities for the 2010 Commonwealth Games (CWGs)—the so-called CWGs Village. These developments have been analyzed elsewhere in more detail (Follmann 2015, 2016a), but it is important to mention here, that in particular the protest against the location of CWGs Village had resulted in a (temporary) no-development moratorium declared by the Lieutenant Governor (LG) of Delhi[2] in September 2007, banning any construction in the River Zone with the exception of the ongoing construction of the CWGs Village and the projects of Delhi Metro.

This chapter focuses on the so-called Millennium Bus Depot developed in the context of the CWGs as it highlights the difficulties associated with the sectoral fragmentation of river governance and missing coordination among state agencies involved in river governance more generally. The case further aims to reveal practices of purification and processes of (institutional) territorialization (Boelens et al. 2016).

Millennium Bus Depot—On or off the Riverbed?

In planning for the 2010 CWGs, the Delhi Transportation Cooperation (DTC) wanted to develop a centrally located temporary bus parking space for the games. Letters between the DTC, the Delhi Development Authority (DDA, the central urban planning authority in Delhi), and the CWGs-Organizing Committee as well as minutes of meetings[3] reveal that three alternative sites for temporary bus parking were discussed of which two (site II and III) fell in the River Zone—Zone O of the Master Plan of Delhi—and one (site I) was located outside the River Zone (see Figure 11.1). While site II was used as a fly-ash dumping area by the Indraprastha power plant, which was known to be shut down in December 2009, site III was used for farming and was affected by seasonal flooding. The land use for the River Zone in this stretch is defined as "River and Water Body" in the Master Plan and "Recreational" in the Zonal Development Plan, respectively. Additionally, site II used to be surrounded by three slums (Sant Nagar, Devi Nagar, and Kali Basti; see Figures 11.2 and 11.3), which had, however, been demolished based on the orders of the High Court, as discussed earlier, and the area was supposed to be given back to the river. In line with these orders and regulations, already in 2008, the DDA—who had initially suggested these sites to the DTC—informed the DTC that in the River Zone "no activity is permitted" and that sites II and III, located in the riverbed, "cannot be developed for the purpose of idle bus parking."[4] Furthermore, different sites for the development of bus depots had been earmarked in the Master

2 The Lieutenant Governor (LG) of Delhi is appointed by the President of India on advice of the Central Government and serves as the constitutional head of the National Capital Territory of Delhi. The LG is further the Chairman of the Delhi Development Authority (DDA). Tejendra Khanna was LG from 9 April 2007 to 9 July 2013.

3 Materials accessed by the NGO Yamuna Jiye Abhiyaan via the RTI Act.

4 Letter DDA to DTC, dated 18 August 2008.

Figure 11.1 Plans for the Yamuna riverbed (ca. 2006) with alternative locations for bus depot (sites I–III) (Source: Materials accessed by YJA via RTI Act. Base map DDA 2006: Draft Zonal Development Plan for the River Zone, demarcation of sites unknown)

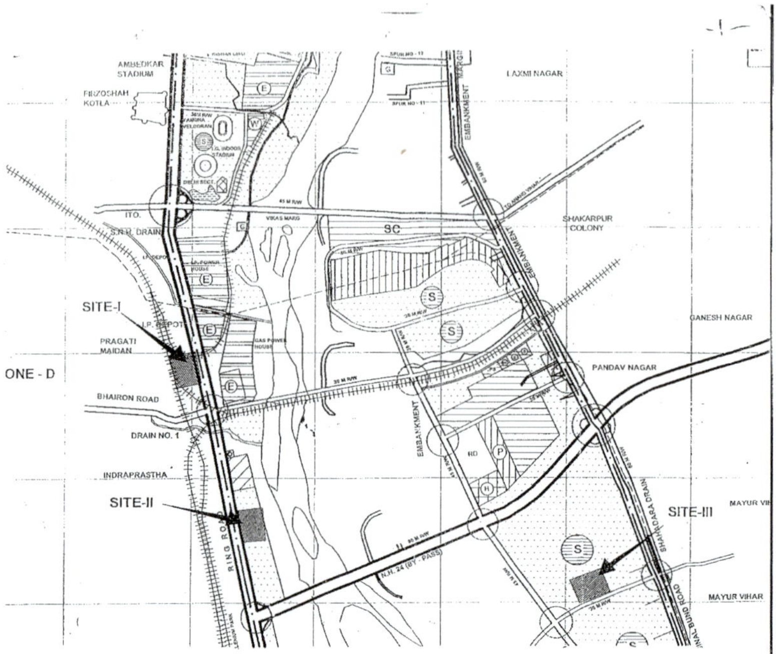

Figure 11.2 DTC Bus Depot in 2012 (Source: Digital Globe, Google Earth)

Figure 11.3 Site of DTC Bus Depot in December 2000 (Source: Digital Globe, Google Earth)

Plan earlier and a parcel of land (6 hectares) right next to the upcoming CWGs Village could have been used for bus parking to transport athletes and officials residing in the CWGs Village (see land already designated for parking (P) in Figure 11.1).

However, the LG allowed the DTC to temporarily use the former fly-ash ponds (site II) for the parking of 300 buses during the CWGs. In the following, the DTC spent about 600 million Indian Rupees (ca. 10 million Euro) and celebrated the construction as "the world's largest bus depot"[5] for the parking of up to 1000 buses (see Figure 11.2).[6] Already in September 2010—shortly before the CWGs—the Transport Minister of the Delhi Government, Arvinder Singh Lovely, was quoted by the media as having said that "After the Games, it will be used as a normal DTC bus depot."[7] Eventually after the CWGs, both the Delhi Government and the DTC—which operates under the Delhi Government—aimed to change the status of the bus depot from temporary to permanent. However, environmental activists—in particular, Manoj Kumar Misra from the NGO Yamuna Jiye Abhiyaan—maintained their protest against the depot and highlighted that the construction of the bus depot was meant to be temporary and violated different laws and

5 https://www.hindustantimes.com/delhi-news/world-s-largest-bus-depot-now-in-city-courtesy-dtc/story-6ILCA05Uagq9PXEe2C80aL.html, accessed 6 May 2018.

6 Construction was carried out by the Delhi-State Public Works Department (PWD).

7 https://www.hindustantimes.com/delhi-news/world-s-largest-bus-depot-now-in-city-courtesy-dtc/story-6ILCA05Uagq9PXEe2C80aL.html, accessed 6 May 2018.

regulations (including the Delhi Master Plan and the Zonal Development Plan for the River Zone) and never received needed clearances, neither in respect to flood control and management (from the Yamuna Standing Committee), urban architectonical guidelines (Delhi Urban Arts Commission), nor general environmental clearance from the Ministry of Environment and Forests. Additionally, the DDA claimed that the land for the depot was never permanently allotted to the DTC.[8]

On these grounds, the activists, Anand Arya and Manoj Kumar Misra, filed a Public Interest Litigation (PIL) in the High Court of Delhi in 2011 requesting that the depot should be demolished and the riverbed restored.[9] In court, the government agencies—in particular the DTC—claimed that the bus depot was constructed to serve a greater public purpose and was not falling on the riverbed. The minutes of a meeting (31 January 2012) of the DDA, the DTC, and the LG reveal that the government agencies tried to align their strategy to make the depot legal: "this land has remained protected from floods by the construction of a bund since the 1960s, i.e., since construction of the IP Power Plant, and has, therefore, no longer the character of river bed/flood-plains [sic]."[10] In the court, the DTC therefore argued that since the land had been earmarked and used for fly-ash dumping for decades it cannot be part of the floodplain anymore. The DTC outlined in an affidavit:

> way back from the 1960s the dumping of fly ash was a permissible activity ... according to the applicable Master Plan. ... The morphology/topography of the land could not subsequently change by itself as to render it part of the river floodplain in the Master Plan for Delhi (MPD) 2021. The river and the riverfront area has since long been sequestered by a bund/flood protection embankment. This segregation was done in the 1960s to protect the Thermal Power Plant from the Yamuna floodplains. Therefore, the land on the west of the bund no longer retains the character of riverbed/floodplain.[11]

Satellite images reveal that earth walls separated a series of fly-ash ponds from each other and from the river; yet the walls towards the river had

8 RTI Response by DDA on 10 November 2010 to NGO Yamuna Jiye Abhiyaan.

9 WP 5481 of 2011 Anand Arya & ano vs Govt of Delhi & others. Another petition was earlier filed by Vinod Kumar Jain against the depot, WP (C) No. 3479/2010 Vinod Kumar Jain vs. GNCTD.

10 Minutes of a meeting held on 31 January 2012 at Raj Niwas "Regarding Bus Depot at Millennium Park."

11 DTC Affidavit, 9 September 2012, in WP 5481 of 2011 Anand Arya & ano vs Govt of Delhi & others.

Figure 11.4 DTC Bus Depot during the floods of 2010 (Photo by Manoj Misra, Yamuna Jiye Abhiyaan [YJA], used with permission, taken on 11 September 2010)

outlets and the overflow from the ponds was released into the river (see Figure 11.3).[12] In 2010 and 2011, the earth wall towards the river was converted into an embankment by the Delhi Irrigation and Flood Department to protect the depot from floods(see Figures 11.4 and 11.5). The construction of the embankment was justified by the need to cater for the adverse flood effects of the new embankment, which was constructed on the other site of the river to protect the CWGs Village (Follmann 2015, 2016a, see Figure 11.1). Notwithstanding these developments, the DDA retained its earlier view that the land of the depot "falls in river front and should be conserved and developed considering the eco-sensitive nature of the River Yamuna."

These different views reveal a political stalemate between the central government (DDA) and the Delhi Government (DTC). This was also reflected in the judgment in March 2013 in which the High Court did not resolve the question of whether the bus depot is on the floodplain or not, but rather

12 The fly ash released from the ponds is the source of arsenic contamination of the groundwater in the Yamuna floodplain above the permissible limit of the World Health Organisation and the Bureau of Indian Standards (Dubey et al. 2012).

Figure 11.5 DTC Bus Depot and new embankment (Photo by Alexander Follmann, taken on 7 November 2011)

emphasized that the construction is not in conformity with the Master Plan. Therefore, the court handed back the decision to the DDA:

> granting six months' time to the respondents to take steps for the change in the Master plan, if it is possible, thereby changing the land use and bringing it in conformity with the present use. In case, the Master Plan is amended in this manner, the natural consequence thereof would be that the Bus Depot would continue to operate from the given site. On the other hand, if attempt to amend the Master plan fails, there would be no [other] option [than] to re-locate the Bus Depot to some other place.[13]

The DDA finally released a public notice for change of land use from "River & Water Body" in the Master Plan and "Recreational" in the Zonal Development Plan to "Transportation" in April 2013—more than seven months after the judgment. Furthermore, the DDA informed the public to change the boundaries of the River Zone in order to exclude the land of the bus depot from

13 HC of Delhi judgment 13 March 2013 in WP 5481 of 2011 Anand Arya & ano vs Govt of Delhi & others.

the River Zone.[14] Against the objections of the public, the DDA in principle approved the DTC's request to change the land use under certain conditions (e.g., soft surface parking, environmental clearance) on 26 July 2013. In the meantime, the National Green Tribunal (NGT)—which was dealing with other cases on the Yamuna in Delhi (Follmann 2016a)—had, however, ordered that "the DDA and other Authorities [sic] shall not act on [the notification] without specific orders of the Tribunal,"[15] and the activists had further gone to the Supreme Court of India. In a landmark judgment on the Yamuna, the NGT had further prohibited any further construction in the River Zone (Follmann 2016a, 311–315).[16] Being eventually a matter in three courts in January 2014, the government agencies met under the chairmanship of the Chief Minister of Delhi and decided that "the DTC would relocate to an alternative site [and] in an overall period of 18 months ... the Millennium Depot would be relocated."[17] After some back and forth regarding new depot sites, the DDA allotted three sites (in total 46.33 hectares) for the relocation of the Millennium Bus Depot in March 2015. Nevertheless, in July 2015, the DTC requested the High Court for another six-month extension to approach the DDA for a change of land use. The High Court dismissed this request and declared that "the steps taken for change of land use in the Master Plan failed" and as "the DDA states that it is not possible for changing the land use in the Master Plan and that alternative lands have also been allotted to the DTC," the Court dismissed the DTCs final plea for the site.[18] In 2016, the Supreme Court also requested the DTC to vacate the site. However, in 2016, the Supreme Court again gave the DTC one more year to obtain a change of land use, since the matter of whether the depot is on the riverbed or not was never resolved.[19] After the buses were shifted in early 2017, the Supreme Court transferred the case to the NGT to avoid further parallel proceedings. An expert committee of the NGT shall now examine whether the site is part of the floodplain or not.

Discussion

The planning, implementation, and post-implementation discourse around the Millennium Bus Depot highlights that powerful state actors have aimed

14 Public Notice DDA File No. F3 (73)/2003-MP/Pt.III, 17 April 2013.
15 NGT 28 October 2012, Application No. 06/2012 Manoj Misra vs. UOI.
16 NGT 28 October 2012, Application No. 06/2012 Manoj Misra vs. UOI.
17 Minutes of meeting 15 January 2014.
18 Order High Court of Delhi, 20 October 2015 in W.P.(C) 5481/2011.
19 Govt. of NCT of Delhi and Anr vs Anand Arya and Ors on 5 February 2016.

and still aim to purify the location of the bus depot as land, not river. The DDA as the central land-use planning authority is generally the actor that draws neat boundaries in the Yamuna riverscape to define the territories of land and water. In the past, this boundary making, however, hardly ever relied on a scientific separation—if we acknowledge that this is possible at all—but rather the river's boundaries as they appear in the land use plans of the DDA were produced materially by the construction of embankments. However, it had been especially the DDA who aimed for large-scale riverfront development in the past, and therefore "has opened up the riverbed for a wide range of uses and, moreover, assured the state's own flexibility in changing the designated land use as needed at a later stage" (Follmann 2015, 220). As outlined above, the DDA had initially suggested the site to the DTC as such a construction was in line with the plans for the reclamation of the riverbed for urban development. Figure 11.1 shows that it was planned (in a draft for the Zonal Development Plan for the River Zone in 2006) to connect the existing embankments and develop a forward embankment along the later CWGs Village crossing the N.H. 24 (By-Pass) Bridge to the south. While these plans were facing increasing opposition, the CWGs, as attested by the central committee interrogating the CWGs-related corruption, "provided a pretext for 'land grab' by various Government agencies after short circuiting the established rules and procedures."[20] Therefore, the (institutional) territorialization (Boelens et al. 2016) around the River Yamuna focused on securing land banks along the river for future development. In this process, conflicts between central government authorities (e.g., the DDA) and Delhi state government authorities reflected the larger dispute about which level governs the city—and the river in the city. With regard to the river, the DTC, in liaison with the Delhi Public Works Department and the Delhi Irrigation and Flood Department, created the material facts in the form of new embankments and the depot. Discursively the DTC followed an earlier argumentation by the DDA that the land beyond the embankment is not part of the riverbed anymore. The DDA had successfully claimed exactly this in the Supreme Court case against the CWGs Village (Follmann 2016a, 263). In doing so, the DDA purified areas behind embankments as land in order to retain them under their unrestrained control as the land-controlling agency. Therefore, practices of purification and processes of (institutional) territorialization come together.

20 High Level Committee for the Commonwealth Games 2011. Commonwealth Games Village – Second Report of HCL. New Delhi. http://www.archive.india.gov.in/high_level/reports.htm, accessed 6 May 2018.

Conclusion

The hybridity of riverscapes produces contradictions for the multiplicity of agencies in river governance, which, based on their sectoral mandates, and in the Indian case often blurred responsibilities, are dependent on inter-sectoral collaboration in order to holistically govern river cities. Urban planners in Delhi—and elsewhere—remain uncomfortable with this hybridity because it makes it difficult or even impossible to fix the boundaries between land and water. Land-use planning through the exercise of master planning, and land-use zoning in particular, institutionalizes the purification of socio-natural hybrids like riverscapes.

Besides these contradictions, the case study of the Millennium Depot on the banks of the Yamuna in Delhi has shown an inconsistency in respect to the environmental protection of the river and its floodplain, which is manifested in a differentiated treatment of slums—seen as a nuisance and polluters—versus so-called planned developments (Follmann 2016a; Ghertner 2011). Therefore, "legality" of land use was not determined by whether it was actually conforming with the existing urban planning or environmental regulations, but rather depended on whether it fit in with the DDA's vision of the riverfront development. The flexible purification of the area along the banks as either land of the river (in the case of the slums) or land beyond the embankment allowed this flexible, differentiated treatment instrumentalizing environmental concerns.

Land use policies (e.g., zoning in the Master and Zonal Development Plans) purify the riverscapes in order to demarcate territories and make the space legible to be governed by the state. The hybridity created contradictions when translated into the water-land binary inherent to land-use planning logics. The case highlights that, as Blomley (2008) argued, "simplification is complicated." While in the past (e.g., in the CWGs Village case), the courts followed the view that embankments mark fixed boundaries between the city and the river (Follmann 2016a), the case of the bus depot revealed that the courts were unwilling—or at least hesitant—to sanction these simplified hydrological logics. Nonetheless, the question whether the depot is legally located on or off the floodplain of the River Yamuna in Delhi remains open. It emerges that the hybridity of riverscapes is alien to environmental jurisprudence; it continues to depend on purification, too. The contradictions associated with the purification of the depot site as land are obvious, especially when the river is "uncooperative" and frequent flooding challenges the neat boundaries—drawn by land use planners and aimed to be fixed by embankments.

Recent urban flood disasters in India (cf., e.g., Arabindoo 2016 and Bremner 2020 for Chennai; Anjaria 2006 for Mumbai) and elsewhere in Asia (cf., e.g., Thanvisitthpon et al. 2018 for Bangkok; Liao, this volume, for Taipei) have highlighted the importance of inter-sectoral coordination in governing urban rivers as well as recalled the need for providing enough space for flood waters in river cities. Overall, the legal battles around the Yamuna in Delhi highlight that techno-managerial approaches to govern urban nature remain often dominant, as powerful actors (in Delhi: the DDA) are able to simplify complex river-city relationships into seemingly neat categories (land/water body), which even stand when they are brought up before court, as legal systems are unable to deal with hybridity either.

The concept of riverscapes is helpful here to acknowledge that (urban) rivers have no neat boundaries and binary conceptualizations of water/land, river/city, and nature/culture will always produce contradictions. Therefore, an understanding of rivers as hybrids is highly needed, and it must find its way into (land-use) planning, in particular in river cities. A possible measure is creating (more) space for the river, long discussed also in India as part of a so-called River Regulation Zone, in line with the existing Coastal Regulation Zone. However, how to integrate hybrid spaces into planning practices and, in particular, how to demarcate the space for the river in legally binding land use plans remain open questions and there is a great need for future research.

Acknowledgments

The author is thankful to Manoj Misra (Yamuna Jiye Abhiyaan) for sharing the information and materials accessed through applications under the Right to Information (RTI) Act. The author would like to thank the organizers and participants of the River Cities Symposium in Surabaya 2017 for their valuable comments. Responsibility for the paper's flaws and limitations must of course remain with the author.

References

Alley, Kelly D. 2012. "The Paradigm Shift in India's River Policies: From Sacred to Transferable Waters." In *Water, Cultural Diversity, and Global Environmental Change: Emerging Trends, Future Sustainability*, edited by B.R. Johnston, Lisa Hiwasaki, Irene J. Klaver, Ameyali Ramos Castillo, and Veronica Strang, 31–48. Heidelberg: Springer.

Anjaria, Jonathan Shapiro. 2006. "Urban Calamities: A View from Mumbai." *Space and Culture*, 9(1): 80–82.

Arabindoo, Pushpa. 2016. "Unprecedented Natures?" *City*, 20(6): 800–821.

Baghel, Ravi. 2014. *River Control in India: Spatial, Governmental and Subjective Dimensions*. Cham: Springer International.

Baud, Isa, and R. Dhanalakshmi. 2007. "Governance in Urban Environmental Management: Comparing Accountability and Performance in Multi-stakeholder Arrangements in South India." *Cities*, 24(2): 133–147.

Baviskar, Amita (ed.). 2007. *Waterscapes: The Cultural Politics of a Natural Resource*. Ranikhet, India: Permanent Black.

Baviskar, Amita. 2011. "What the Eye Does Not See: The Yamuna in the Imagination of Delhi." *Economic and Political Weekly*, 46(50): 45–53.

Bhan, Gautam. 2009. "'This Is No Longer the City I Once Knew': Evictions, the Urban Poor and the Right to the City in Millennial Delhi." *Environment and Urbanization*, 21(1): 127–142.

Billé, Raphaël. 2008. "Integrated Coastal Zone Management: Four Entrenched Illusions." *Surveys and Perspectives Integrating Environment and Society*, 1(2): 7–86.

Blomley, Nicholas. 2008. "Simplification Is Complicated: Property, Nature, and the Rivers of Law." *Environment and Planning A*, 40(8): 1825–1842.

Boelens, Rutgerd, Jaime Hoogesteger, Erik Swyngedouw, Jeroen Vos, and Philippus Wester. 2016. "Hydrosocial Territories: A Political Ecology Perspective." *Water International*, 41(1): 1–14.

Bremner, Lindsay. 2020. "Planning the 2015 Chennai Floods." *Environment and Planning E: Nature and Space*, 3(2): 732–760.

Bryant, Raymond L., and Sinéad Bailey. 1997. *Third World Political Ecology*. London: Routledge.

Coelho, Karen, and Nithya Raman. 2010. "Salvaging and Scapegoating: Slum Evictions on Chennai's Waterways." *Economic and Political Weekly*, 45(21): 19–23.

Coelho, Karen, and Nithya V. Raman. 2013. "From the Frying-Pan to the Floodplain: Negotiating Land, Water and Fire in Chennai's Development." In *Ecologies of Urbanism in India: Metropolitan Civility and Sustainability*, edited by Anne Rademacher and K. Sivaramakrishnan, 145–168. Hong Kong: Hong Kong University Press.

D'Souza, Rohan. 2006. *Drowned and Dammed: Colonial Capitalism and Flood Control in Eastern India*. New Delhi: Oxford University Press.

Desai, Renu. 2012. "Governing the Urban Poor: Riverfront Development, Slum Resettlement and the Politics of Inclusion in Ahmedabad." *Economic and Political Weekly*, 47(2): 49–56.

Desfor, Gene, Jennifer Laidley, Quentin Stevens, and Dirk Schubert (eds.). 2011. *Transforming Urban Waterfronts: Fixity and Flow*. New York, NY: Routledge.

Dubey, Chandra, Bhupendra Mishra, Dericks Shukla, Ravindra Singh, M. Tajbakhsh, and Purushottam Sakhare. 2012. "Anthropogenic Arsenic Menace in Delhi Yamuna Flood Plains." *Environmental Earth Sciences*, 65(1): 131–139.

Follmann, Alexander. 2015. "Urban Mega-projects for a 'World-Class' Riverfront: The Interplay of Informality, Flexibility and Exceptionality along the Yamuna in Delhi, India." *Habitat International*, 45(3): 213–222.

Follmann, Alexander. 2016a. *Governing Riverscapes: Urban Environmental Change along the River Yamuna in Delhi, India*. Stuttgart: Steiner.

Follmann, Alexander. 2016b. "The Role of Environmental Activists in Governing Riverscapes: The Case of the Yamuna in Delhi, India." *South Asia Multidisciplinary Academic Journal* (SAMAJ) [Online] 2016(14). https://doi.org/10.4000/samaj.4184.

Forsyth, Tim. 2003. *Critical Political Ecology: The Politics of Environmental Science*. London: Routledge.

Ghertner, D. Asher. 2011. "Green Evictions: Environmental Discourse of a 'Slum-Free' Delhi." In *Global Political Ecology*, edited by Richard Peet, Paul Robbins, and Michael Watts, 145–165. London: Routledge.

Haberman, David L. 2006. *River of Love in an Age of Pollution: The Yamuna River of Northern India*. Berkeley, CA: University of California Press.

Haidvogl, Gertrud. 2012. "The Channelization of the Danube and Urban Spatial Development in Vienna in the Nineteenth and Early Twentieth Centuries." In *Urban Rivers: Remaking Rivers, Cities, and Space in Europe and North America*, edited by Stéphane Castonguay and Matthew Evenden, 113–129. Pittsburgh, PA: University of Pittsburgh Press.

Hartmann, Thomas. 2013. "Land Policy for German Rivers: Making Space for the Rivers." In *Making Space for the River: Governance Experiences with Multifunctional River Flood Management in the US and Europe*, edited by Jeroen Frank Warner, Arwin van Buuren, and Jurian Edelenbos, 121–134. London: IWA.

Hartmann, Thomas, and Tejo Spit. 2014. "Frontiers of Land and Water Governance in Urban Regions." *Water International*, 39(6): 791–797.

Hirsch, Philip. 2006. "Water Governance Reform and Catchment Management in the Mekong Region." *The Journal of Environment and Development*, 15(2): 184–201.

Hohn, Uta, and Birgit Neuer. 2006. "New Urban Governance: Institutional Change and Consequences for Urban Development." *European Planning Studies*, 14(3): 291–298.

Holifield, Ryan, and Nick Schuelke. 2015. "The Place and Time of the Political in Urban Political Ecology: Contested Imaginations of a River's Future." *Annals of the Association of American Geographers*, 105(2): 294–303.

Iyer, Ramaswamy R. (ed.). 2015. *Living Rivers, Dying Rivers*. New Delhi: Oxford University Press.

Karpouzoglou, Timothy, and Sumit Vij. 2017. "Waterscape: A Perspective for Understanding the Contested Geography of Water." *Wiley Interdisciplinary Reviews: Water*, 4(3): e1210.

Karpouzoglou, Timothy, and Anna Zimmer. 2016. "Ways of Knowing the Wastewaterscape: Urban Political Ecology and the Politics of Wastewater in Delhi, India." *Habitat International*, 54(2): 150–160.

Keeley, James, and Ian Scoones. 2003. *Understanding Environmental Policy Processes: Cases from Africa*. London: Earthscan.

Lahiri-Dutt, Kuntala. 2014. "Beyond the Water-Land Binary in Geography: Water/Lands of Bengal Re-visioning Hybridity." *ACME*, 13(3): 505–529.

Lahiri-Dutt, Kuntala, and Gopa Samanta. 2013. *Dancing with the River: People and Life on the Chars of South Asia*. New Haven, CT and London: Yale University Press.

Latour, Bruno. 1993. *We Have Never Been Modern*. London: Harvester Wheatsheaf.

Latour, Bruno. 2004. *Politics of Nature: How to Bring Science into Democracy*. Cambridge, MA: Harvard University Press.

Lawhon, Mary, Henrik Ernstson, and Jonathan Silver. 2014. "Provincializing Urban Political Ecology: Towards a Situated UPE through African Urbanism." *Antipode*, 46(2): 497–516.

Loftus, Alex. 2012. *Everyday Environmentalism: Creating an Urban Political Ecology*. Minneapolis, MN: University of Minnesota Press.

Lübken, Uwe. 2012. "Rivers and Risk in the City: The Urban Floodplain as a Contested Space." In *Urban Rivers: Remaking Rivers, Cities, and Space in Europe and North America*, edited by Stéphane Castonguay and Matthew Evenden, 130–144. Pittsburgh, PA: University of Pittsburgh Press.

Maria, Augustin. 2008. *Urban Water Crisis in Delhi: Stakeholder Responses and Potential Scenarios of Evolution*. Paris, France: IDDRI.

Murdoch, Jonathan. 2006. *Post-structuralist Geography: A Guide to Relational Space*. London: Sage.

Myers, Garth Andrew. 2008. "Sustainable Development and Environmental Justice in African Cities." *Geography Compass*, 2(3): 695–708.

Pessina, Gloria. 2019. "The Brand-New Riverfront and the Historical Centre: Narratives and Open Questions in Contemporary Ahmedabad, India." *European Journal of Creative Practices in Cities and Landscapes*, 2(1): 65–92.

Rademacher, Anne M. 2011. *Reigning the River: Urban Ecologies and Political Transformation in Kathmandu*. Durham, NC: Duke University Press.

Scott, James C. 1998. *Seeing Like a State: How Certain Schemes to Improve the Human Condition Have Failed*. New Haven, CT: Yale University Press.

Sharan, Awadhendra. 2016. "A River and the Riverfront: Delhi's Yamuna as an In-Between Space." *City, Culture and Society*, 7(4): 267–273.

Smith, Neil. 1984. *Uneven Development: Nature, Capital and the Production of Space*. Oxford: Blackwell.

Swyngedouw, Erik. 1999. "Modernity and Hybridity: Nature, Regeneracionismo, and the Production of the Spanish Waterscape, 1890–1930." *Annals of the Association of American Geographers*, 89(3): 443–465.

Swyngedouw, Erik. 2004. *Social Power and the Urbanization of Water*. Oxford: Oxford University Press.

Thanvisitthpon, Nawhath, Sangam Shrestha, and Indrajit Pal. 2018. "Urban Flooding and Climate Change: A Case Study of Bangkok, Thailand." *Environment and Urbanization ASIA*, 9(1): 86–100.

Whatmore, Sarah. 2002. *Hybrid Geographies: Natures, Cultures, Spaces*. London: Sage.

Whatmore, Sarah, and Susan Boucher. 1993. "Bargaining with Nature: The Discourse and Practice of 'Environmental Planning Gain'." *Transactions of the Institute of British Geographers*, 18(2): 166–178.

Wiering, Mark, and Irene Immink. 2006. "When Water Management Meets Spatial Planning: A Policy-Arrangements Perspective." *Environment and Planning C: Government and Policy*, 24(3): 423–438.

Zérah, Marie-Hélène. 2009. "Participatory Governance in Urban Management and the Shifting Geometry of Power in Mumbai." *Development and Change*, 40(5): 853–877.

Biography

Dr. Alexander Follmann studied geography in Cologne (Germany) and Vancouver (Canada). His PhD research (University of Cologne) on the River Yamuna in Delhi was funded by PhD fellowships from the German Academic Exchange Service (DAAD) and the German Academic Scholarship Foundation. He currently works at the Institute of Geography, University of Cologne.

Index

For Product Safety Concerns and Information please contact our EU representative GPSR@taylorandfrancis.com Taylor & Francis Verlag GmbH, Kaufingerstraße 24, 80331 München, Germany